NATEF Standards Job Sheets

Heating and Air Conditioning (A7)

Second Edition

Jack Erjavec

THOMSON
DELMAR LEARNING™

Australia Canada Mexico Singapore Spain United Kingdom United States

NATEF Standards Job Sheets

Heating and Air Conditioning (A7)
Second Edition

Jack Erjavec

Vice President, Technology and Trades SBU:
Alar Elken

Editorial Editor:
Sandy Clark

Senior Acquisitions Editor:
David Boelio

Development Editor:
Matthew Thouin

Marketing Director:
David Garza

Channel Manager:
William Lawrensen

Marketing Coordinator:
Mark Pierro

Production Director:
Mary Ellen Black

Production Editor:
Toni Hansen

Art/Design Specialist:
Cheri Plasse

Technology Project Manager:
Kevin Smith

Editorial Assistant:
Andrea Domkowski

Printed in United States of America
2 3 4 5 XXXX 09 08 07 06

For more information contact Thomson Delmar Learning, Executive Woods, 5 Maxwell Drive, Clifton Park, NY 12065-2919.
Or find us on the World Wide Web at http://www.delmarlearning.com.

ISBN-13: 978-1-4180-2080-4
ISBN-10: 1-4180-2080-X

NOTICE TO THE READER

CONTENTS

PREFACE

The automotive service industry continues to change with the technological changes made by automobile and tool and equipment manufacturers. Today's automotive technician must have a thorough knowledge of automotive systems and components, good computer skills, exceptional communication skills, good reasoning, the ability to read and follow instructions, and above average mechanical aptitude and manual dexterity.

This new edition, like the last, was designed to give students a chance to develop the same skills and gain the same knowledge that today's successful technician has. This edition also reflects the changes in the guidelines established by the National Automotive Technicians Education Foundation (NATEF), as of July 2005.

The purpose of NATEF is to evaluate technician training programs against standards developed by the automotive industry and recommend qualifying programs for certification (accreditation) by ASE (National Institute for Automotive Service Excellence). Programs can earn ASE certification upon the recommendation of NATEF. NATEF's national standards reflect the skills that students must master. ASE certification through NATEF evaluation ensures that certified training programs meet or exceed industry-recognized, uniform standards of excellence.

At the expense of much time and many minds, NATEF has assembled a list of basic tasks for each of their certification areas. These tasks identify the basic skills and knowledge levels that competent technicians have. The tasks also identify what is required for a student to start a successful career as a technician.

Most of the content in this book are job sheets. These job sheets relate to the tasks specified by NATEF. The main considerations during the creation of these job sheets were student learning and program certification by NATEF. Students are guided through standard industry accepted procedures. While they are progressing, they are asked to report their findings as well as offer their thoughts on the steps they have just completed. The questions asked of the students are thought provoking and require students to apply what they know to what they observe.

The job sheets were also designed to be generic. That is, whenever possible, the tasks can be performed on any vehicle from any manufacturer. Also, completion of the sheets does not require the use of specific brands of tools and equipment; rather students use what is available. In addition, the job sheets can be used as a supplement to any good textbook.

Also included are description and basic use of the tools and equipment listed in NATEF's standards. The standards recognize that not all programs have the same needs, nor do all programs teach all of the NATEF tasks. Therefore, the basic philosophy for the tools and equipment requirement is that the training should be as thorough as possible with the tools and equipment necessary for those tasks.

Theory instruction and hands-on experience of the basic tasks provide initial training for employment in automotive service or further training in any or all of the specialty areas. Competency in the tasks indicates to employers that you are skilled in that area. You need to know the appropriate theory, safety, and support information for each required task. This should include identification and use of the required tools and testing and measurement equipment required for the tasks, the use of current reference and training materials, the proper way to write work orders and warranty reports, and the storage, handling, and use of Hazardous Materials as required by the 'Right to Know Law', and federal, state, and local governments.

Words to the Instructor: I suggest you grade these job sheets based on completion and reasoning. Make sure the students answer all questions. Then look at their reasoning to see if the task was actually completed and to get a feel for their understanding of the topic. It will be easy for students to copy others' measurements and findings, but each student should have their own base of understanding and that will be reflected in their explanations.

Words to the Student: While completing the job sheets, you have a chance to develop the skills you need to be successful. When asked for your thoughts or opinions, think about what you observed. Think about what could have caused those results or conditions. You are not being asked to give accurate explanations for everything you do or observe. You are only asked to think. Thinking leads to understanding. Good technicians are good because they have a basic understanding of what they are doing and of why they are doing it.

Jack Erjavec

HEATING AND AIR CONDITIONING SYSTEMS

To prepare you to learn what you should learn from completing the job sheets, some basics must be covered. This discussion begins with an overview of heating and air conditioning systems. Emphasis is placed on what they do and how they work, including the major components of heating and air conditioning systems and their role in the efficient operation of heating and air conditioning systems of all designs.

Preparation for work on an automobile would not be complete without addressing certain safety issues. This discussion covers what you should and should not do while working on heating and air conditioning systems, including the proper ways to deal with hazardous and toxic materials.

NATEF's task list for Heating and Air Conditioning certification is given, with definitions of some terms used to describe the tasks. This list gives you a good look at what the experts say you need to know before you can be considered competent to work on heating and air conditioning systems.

Following the task list are descriptions of the various tools and types of equipment you must be familiar with. These are the tools you will use to complete the job sheets. They are also the tools NATEF has identified as necessary for servicing heating and air conditioning systems.

After the tool discussion is a cross-reference guide that shows which NATEF tasks are related to specific job sheets. In most cases, there are single job sheets for each task. Some tasks are part of a procedure, in which case one job sheet may cover two or more tasks. The remainder of the book contains the job sheets.

HEATING SYSTEMS

The main components of an automotive heating system are the heater core, the heater control valve, the blower motor and the fan, and the heater and defroster ducts. The heating system works with the engine's cooling system and converts the heat from the coolant circulating inside the engine to hot air, which is blown into the passenger compartment. A heater hose transfers hot coolant from the engine to the heater control valve and from there to the heater core inlet. As the coolant circulates through the core, heat is transferred from the coolant to the tubes and fins of the core. Air blown through the core by the blower motor and fan then picks up heat from the surfaces of the core and transfers it into the passenger compartment. After giving up its heat, the coolant is then pumped out through the heater core outlet, where it is returned to the engine's cooling system to be heated again.

The heater control valve (also called the water flow valve) controls the flow of coolant into the heater core from the engine. In a closed position, the valve allows no flow of hot coolant to the heater core, thereby keeping the heater core cool. In an open position, the valve allows heated coolant to circulate through the heater core, thereby maximizing heater efficiency.

Cable-operated heater control valves are controlled directly by the heater control lever on the dashboard. Thermostatically controlled valves have a liquid-filled capillary tube located in the discharge air stream off the heater core. This tube senses air temperature, and the valve modulates the flow of water to maintain a constant temperature, regardless of engine speed or temperature.

Most heater control valves are vacuum operated. When a vacuum signal reaches the valve, a diaphragm inside the valve is raised, either opening or closing the valve against an opposing spring. When the temperature selection on the heater control panel is changed, vacuum to the valve is vented and the valve returns to its original position. Some vehicles don't use a heater control valve; instead, a heater door controls how much heat is released into the passenger compartment from the heater core.

The blower motor is usually located in the heater housing assembly. A multiposition switch on the control panel controls its speed. The switch works in connection with a stepped resistor block normally located near the heater housing. The resistor block controls blower motor speed. The typical resistor block has three or four resistors in series with the blower motor. The multiposition switch directs voltage to different locations in this series circuit. When low fan speed is selected, voltage is present at the beginning of the series circuit. High resistance causes slower operation of the blower motor. When high speed is selected, voltage is present at the end, or near the end, of the resistor assembly. Low resistance allows for higher fan speeds. On some vehicles, when the engine is running, the blower motor runs constantly at low speed.

Transferring heated air from the heater core to the passenger compartment is the job of the heater and defroster ducts. The ducts are typically part of a large plastic shell that connects to the necessary inside and outside vents. This ductwork also has mounting points for the evaporator and heater core assemblies. Also contained inside the duct are the doors required to direct air to the floor, dash, and/or windshield. Sometimes the duct is connected directly to the vents; other times, hoses are used.

Ventilation System

The ventilation system is designed to supply outside air to the passenger compartment through upper or lower vents, or both. Several systems are used to vent air into the passenger compartment. The most common is the flow-through system. In this arrangement, a supply of outside air, which is called *ram air,* flows into the car when it is moving. When the car is not moving, the heater fan can produce a steady flow of outside air. In operation, air is circulated throughout the passenger compartment. From there the air is forced outside the vehicle through an exhaust area.

AIR CONDITIONING SYSTEMS

It is important to know the basic theory of automotive air conditioning systems. Understanding how an A/C system moves heat from the confined space of the passenger compartment and dissipates it into the atmosphere will help you diagnose A/C problems.

An air conditioning system (Figure 1) is designed to pump heat from one point to another. All materials or substances, even those as cold as –459° Fahrenheit, have heat in them. Also, heat always flows from a warmer object to a colder one. The greater the temperature difference between the objects, the greater the amount of heat flow.

Objects can be in one of three forms or states: solid, liquid, or gas. When objects change from one state to another, large amounts of heat can be transferred. For example, when water temperature goes below 32°F, water changes from a liquid to a solid (ice). If the temperature of water is raised to 212°F, the liquid turns into a gas (steam). An interesting thing occurs when water, or any matter, changes from a solid to a liquid and then from a liquid to a gas. Additional heat is necessary to change the state of the substance, even though this heat does not register on a thermometer. For example, ice at 32°F requires heat to change into water, which will also be at 32°F. Additional heat raises the temperature of the water until it reaches the boiling point of 212°F. More heat is required to change water into steam. If the temperature of the steam were measured, however, it would also be 212°F. The amount of heat necessary to change the state of a substance is called *latent heat*—or *hidden heat*—because it cannot be measured with a thermometer. This hidden heat is the basic principle behind all air conditioning systems.

Pressure on a substance, such as a liquid, changes its boiling point. The greater the pressure on a liquid, the higher the boiling point. If pressure is placed on a vapor, the vapor condenses at

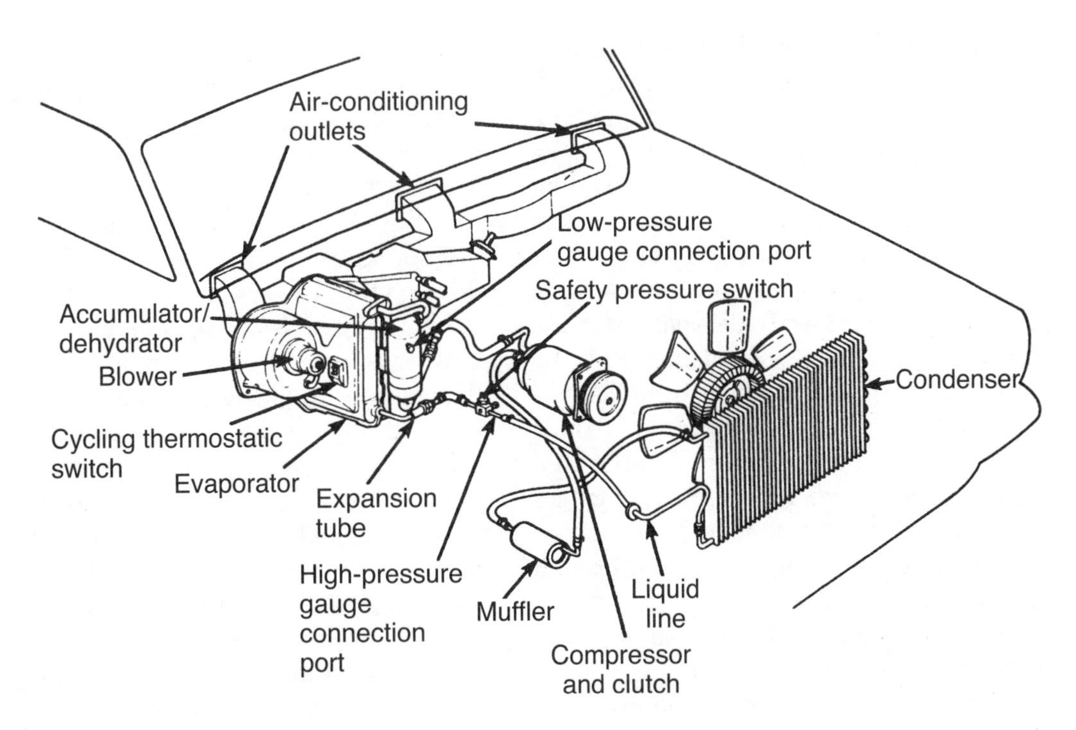

Figure 1 A typical air conditioning system.

a higher-than-normal temperature. In addition, as the pressure on a substance is reduced, the boiling point can also be reduced. For example, the boiling point of water is 212°F. Increasing the pressure on the fluid can increase the boiling point. Reducing the pressure or placing the fluid in a vacuum can decrease it.

Refrigerants

The substance used to remove heat from the inside of an air-conditioned vehicle is called the refrigerant. On older automotive air conditioning systems, the refrigerant was Refrigerant-CFC-12 (commonly referred to as R-12 and Freon). R-12 is dichlorodifluoromethane (CCl_2F_2). By law, R-12 is no longer used in A/C systems. It was found that the chemicals in R-12 were damaging the earth's ozone layer, which is the earth's outermost shield of protection against the harmful effects of the sun's ultraviolet rays. The ozone depletion is caused, in part, by release of chlorofluorocarbons (CFCs) into the atmosphere. R-12 is in the chemical family of CFCs. Since air conditioning systems with R-12 are susceptible to leaks, further damage to the ozone layer could be avoided by stopping the use of R-12 in air conditioning units.

To replace R-12 in A/C systems, the manufacturers have agreed to use R-134a. This refrigerant may also be referred to as SUVA. R-134a is a hydrofluorocarbon (HFC) that causes less damage to the ozone layer when released to the atmosphere.

Although R-134a air conditioners operate in the same way and with the same basic components as R-12 systems, the two refrigerants are not interchangeable. Because it is less efficient than R-12, R-134a must operate at higher pressures. Therefore, R-134a systems use components designed for those higher pressures. To keep systems operating that were originally equipped with R-12, there are retrofit kits available for a changeover to R-134a. Although R-134a is less likely to have an adverse effect on the ozone layer, it still has the capability of contributing to the greenhouse effect when released into the air. Therefore, the recovery and recycling of R-12 and R-134a is mandatory by law. Laws have also been passed that dictate the certification of equipment and technicians.

Basic Operation

Refrigerants are used to carry heat from the inside of the vehicle to outside the vehicle. Automotive refrigerants have a low boiling point (the point at which evaporation occurs). As a refrigerant changes state, it absorbs a large amount of heat. Since the heat that it absorbs is from the inside of the vehicle, passengers become cooler.

To understand how a refrigerant is used to cool the interior of a vehicle, the effects of pressure and temperature on it must be understood. If the pressure of the refrigerant is high, so is its temperature. If the pressure is low, so is its temperature. Therefore, changing its pressure can change the temperature of a refrigerant.

To absorb heat, the temperature and pressure of the refrigerant is kept low. To dissipate heat, the temperature and pressure is kept high. As the refrigerant absorbs heat, it changes from a liquid to a vapor. As it dissipates heat, it changes from a vapor to a liquid. The change from a vapor to a liquid is called *condensation*. These two changes of state—evaporation and condensation—occur continuously as the refrigerant circulates through the air conditioning system.

An A/C system is a closed, pressurized system. It consists of a compressor, condenser, receiver/drier or accumulator, expansion valve or orifice tube, and an evaporator. In a basic air conditioning system, the heat is absorbed and transferred in the following steps.

1. Refrigerant leaves the compressor as a high-pressure, high-temperature vapor.
2. By removing heat via the condenser, the vapor becomes a high-pressure, high-temperature liquid.
3. Moisture and contaminants are removed by the receiver/drier, where the cleaned refrigerant is stored until it is needed.
4. The expansion valve controls the flow of refrigerant into the evaporator.
5. Heat is absorbed from the air inside the passenger compartment by the low-pressure, warm refrigerant, causing the liquid to vaporize and greatly decrease its temperature.
6. The refrigerant returns to the compressor as a low-pressure, low-temperature vapor.

The best way to understand the purpose of an A/C system's major components is to divide the systems into two sides: the high side and the low side. The high side is under high pressure and high temperature. The low side is the low-pressure, low-temperature side of the system.

The Compressor

The compressor separates the high-pressure and low-pressure sides of the system. Its primary purpose is to draw the low-pressure and low-temperature vapor from the evaporator and compress this vapor into high-temperature, high-pressure vapor. This action results in the refrigerant having a higher temperature than surrounding air and enables the condenser to condense the vapor back to a liquid. The secondary purpose of the compressor is to circulate or pump the refrigerant through the A/C system under the different pressures required for proper operation. The compressor is located on the engine and is driven by the engine's crankshaft via a drive belt.

Although many types of compressors are in use today, they are usually one of three types. A *piston compressor* can have its pistons arranged in an in-line, axial, radial, or V design. It is designed to have an intake stroke and a compression stroke for each cylinder. On the intake stroke, the refrigerant from the low side of the system (evaporator) is drawn into the compressor. Refrigerant is taken in through intake reed valves, which are one-way valves that control the flow of refrigerant vapors into the cylinder. During the compression stroke, the vaporous refrigerant is compressed. This increases both the pressure and the temperature of the heat-carrying refrigerant. The outlet or discharge-side reed valves then open to allow the refrigerant to move to the condenser. The outlet reed valves are the beginning of the high side of the system.

A common variation of a piston-type compressor is the variable displacement compressor. These compressors not only act as a regular compressor, but they also control the amount of refrigerant that passes through the evaporator. The pistons are connected to a wobble-plate. The angle of the wobble-plate determines the stroke of the pistons and is controlled by the difference in pressure between the compressor's outlet and inlet. When the stroke of the pistons increases, more refrigerant is being pumped and there is increased cooling.

A *rotary vane compressor* doesn't have pistons. Instead, it consists of a rotor with several vanes and a carefully shaped housing. As the compressor shaft rotates, the vanes and housing form chambers. The refrigerant is drawn through the suction port into these chambers, which become smaller as the rotor turns. The discharge port is located at the point where the gas is completely compressed.

The *scroll-type compressor* has a movable scroll and a fixed scroll that provide an eccentric-like motion. As the compressor's crankshaft rotates, the movable scroll forces the refrigerant against the fixed scroll and toward the center of the compressor. This motion pressurizes the refrigerant. The pressure of air moving in a circular pattern increases as it moves toward the center of the circle. A delivery port is positioned at the center of the compressor and allows the high-pressure refrigerant to flow into the air conditioning system.

Refrigerant Oils

Normally the only source of lubrication for a compressor is the oil mixed with the refrigerant. Because of the loads and speeds at which the compressor operates, proper lubrication is a must for long compressor life. The refrigerant oil required by the system depends on a number of things, but it is primarily dictated by the refrigerant used in the system. R-12 systems use a mineral oil. Mineral oil mixes well with R-12 without breaking down. Mineral oil, however, cannot be used with R-134a. R-134a systems require a synthetic oil, poly alkaline glycol (PAG). There are a number of different blends of PAG oil; always use the one recommended by the vehicle manufacturer or compressor manufacturer.

Compressor Clutches

Compressors are equipped with an electromagnetic clutch as part of the compressor pulley assembly. The clutch is designed to engage the pulley to the compressor shaft when the clutch coil is energized. The clutch provides a way for turning the compressor on or off.

The clutch is driven by power from the engine's crankshaft. When the clutch is engaged, power is transmitted from the pulley to the compressor shaft by the clutch drive plate. When the clutch is not engaged, the compressor shaft does not rotate, and the pulley freewheels.

The clutch is engaged by a magnetic field and disengaged by springs when the magnetic field is broken. When the controls call for compressor operation, the electrical circuit to the clutch is completed, the magnetic clutch is energized, and the clutch engages the compressor. When the electrical circuit is opened, the clutch disengages the compressor.

Condenser

The condenser consists of coiled refrigerant tubing mounted in a series of thin cooling fins to provide maximum heat transfer in a minimum amount of space. The condenser is normally mounted just in front of the vehicle's radiator. It receives the full flow of ram air from the movement of the vehicle or airflow from the radiator fan when the vehicle is standing still.

The condenser condenses or liquefies the high-pressure, high-temperature vapor coming from the compressor. To do so, it must give up its heat. The condenser receives very hot (normally 200° to 400°F), high-pressure refrigerant vapor from the compressor through its discharge hose. The refrigerant vapor enters the inlet at the top of the condenser, and as the hot vapor passes down through the condenser coils, heat moves from the hot refrigerant into the cooler air as it flows across the condenser coils and fins. This process causes a large quantity of heat to be transferred to the outside air and the refrigerant to change from a high-pressure hot vapor to a high-pressure warm liquid. This high-pressure warm liquid flows from the outlet at the bottom of the condenser through a line to the receiver/drier, or to the refrigerant metering device if an accumulator instead of a drier is used.

Receiver/Drier

The receiver/drier is a storage tank for the liquid refrigerant from the condenser. The refrigerant flows into the upper portion of the receiver tank, which contains a bag of desiccant (moisture-absorbing material, such as silica alumina or silica gel). The desiccant absorbs unwanted water and moisture in the refrigerant. As the refrigerant flows through the outlet at the lower portion of the receiver, a mesh screen filters it.

Included in many receiver/driers are a high-pressure fitting, a pressure relief valve, and a sight glass for determining the state and condition of the refrigerant in the system.

Accumulator

Most late-model systems are equipped with an accumulator rather than a receiver/drier. The accumulator is connected into the low side, at the outlet of the evaporator. The accumulator also contains a desiccant and is designed to store excess refrigerant and to filter and dry the refrigerant. If liquid refrigerant flows out of the evaporator, it will be collected by and stored in the accumulator. The main purpose of an accumulator is to prevent liquid from entering the compressor.

Thermostatic Expansion Valve or Orifice Tube

The refrigerant flow to the evaporator must be controlled to obtain maximum cooling, while ensuring complete evaporation of the liquid refrigerant within the evaporator. This is the job of a thermostatic expansion valve (TEV or TXV) or a fixed orifice tube.

The TEV is mounted at the inlet to the evaporator and separates the high-pressure side of the system from the low-pressure side. The TEV regulates refrigerant flow to the evaporator to prevent evaporator flooding or starving. In operation, the TEV regulates the refrigerant flow to the evaporator by balancing the inlet flow to the outlet temperature.

Both externally and internally equalized TEVs are used in air conditioning systems. The only difference between the two valves is that the external TEV uses an equalizer line connected to the evaporator outlet line as a means of sensing evaporator outlet pressure. The internal TEV senses evaporator inlet pressure through an internal equalizer passage. Both valves have a capillary tube to sense evaporator outlet temperature.

Like the TEV, the orifice tube is the dividing point between the high- and low-pressure parts of the system. However, its metering or flow rate control does not depend on comparing evaporator pressure and temperature. It is a fixed orifice. The flow rate is determined by pressure difference across the orifice and by subcooling. *Subcooling* is additional cooling of the refrigerant in the bottom of the condenser after it has changed from vapor to liquid. The flow rate through the orifice is more sensitive to subcooling than to pressure difference.

Evaporator

The evaporator, like the condenser, consists of a refrigerant coil mounted in a series of thin cooling fins. It provides a maximum amount of heat transfer in a minimum amount of space. The evaporator is usually located beneath the dashboard or instrument panel.

On receiving the low-pressure, low-temperature liquid refrigerant from the thermostatic expansion valve or orifice tube in the form of an atomized (or droplet) spray, the evaporator serves as a boiler or vaporizer. The regulated flow of refrigerant boils immediately. Heat from the core surface is lost to the boiling and vaporizing refrigerant, which is cooler than the core, thereby cooling the core. The air passing over the evaporator loses its heat to the cooler surface of the core, thereby cooling the air inside the car. As the process of heat loss from air to the evaporator core surface is taking place, any moisture (humidity) in the air condenses on the outside of the evaporator core and is drained off as water. This dehumidification of air adds to passenger comfort. It can also be used to control the fogging of windows.

Through the metering, or controlling, action of the thermostatic expansion valve or orifice tube, greater or lesser amounts of refrigerant are provided in the evaporator to adequately cool the car under all heat load conditions. If too much refrigerant is allowed to enter, the evaporator floods. This results in poor cooling due to the higher pressure (and temperature) of the refrigerant. The refrigerant can neither boil away rapidly nor vaporize. On the other hand, if too little refrigerant is metered, the evaporator starves. Poor cooling again results because the refrigerant boils away or vaporizes too quickly before passing through the evaporator.

Refrigerant Lines

All the major components of the system have inlet and outlet connections that accommodate either flare or O-ring fittings. There are three major refrigerant lines. Suction lines are located between the outlet side of the evaporator and the inlet side or suction side of the compressor. They carry the low-pressure, low-temperature refrigerant vapor to the compressor, where it again is recycled through the system. Suction lines are always distinguished from the discharge lines by touch and size. They are cold to the touch. The suction line is larger in diameter than the liquid line because refrigerant in a vapor state takes up more room than refrigerant in a liquid state.

The discharge or high-pressure line connects the compressor to the condenser. The liquid lines connect the condenser to the receiver/drier and the receiver/drier to the inlet side of the expansion valve. Through these lines, the refrigerant travels in its path from a gas state (compressor outlet) to a liquid state (condenser outlet) and then to the inlet side of the expansion valve, where it vaporizes on entry to the evaporator. Discharge and liquid lines are always very warm to the touch and easily distinguishable from the suction lines.

Aluminum tubing is commonly used to connect air conditioning components where flexibility is not required. R-134a systems are required to be fitted with quick-disconnect fittings through the system. These also have hoses specially made for R-134a. They have an additional layer of rubber that serves as a barrier to prevent the refrigerant from escaping through the pores of the hose.

Sight Glass

The sight glass allows the flow of refrigerant in the lines to be observed. A sight glass is normally found on systems using R-12 and a thermal expansion valve. It is not commonly found on R-134a systems. It can be located on the receiver/drier or in-line between the receiver/drier and the expansion valve or tube.

Accumulator systems do not have a receiver/drier to separate the gas from the liquid as it flows from the condenser. The liquid line always contains a certain amount of bubbles, therefore, it would be useless to have a sight glass in these systems. Pressure and performance testing are the only ways to identify low refrigerant levels.

Blower Motor and Fan

The blower motor/fan assembly is located in the evaporator housing. The blower, which is basically the same type as those used in heater systems, draws warm air from the passenger compartment, forces it over the coils and fins of the evaporator, and blows the cooled, cleaned, and dehumidified air into the passenger compartment. A fan switch controls the blower motor. During cold weather, the blower motor/fan assembly provides the airflow, and the heater core provides the heat for the passenger compartment.

Evaporator Pressure Control Systems

Evaporator controls maintain back pressure in the evaporator. Because of the refrigerant temperature-pressure relationship, the effect is to regulate evaporator temperature. The temperature is controlled to a point that provides effective air cooling but prevents the freezing of moisture that condenses on the evaporator.

In this type of system, the compressor operates continually when dash controls are in the air conditioning position. An evaporator pressure control valve automatically controls evaporator outlet air temperature. These types of valves throttle the flow of refrigerant out of the evaporator as required to establish a minimum evaporator pressure and thereby prevent freezing of condensation on the evaporator core.

Cycling Clutch Systems

In cycling clutch systems, the compressor is run intermittently by controlling the application and release of its clutch by a thermostatic or pressure switch. The thermostatic switch senses the evaporator's outlet air temperature through a capillary tube that is part of the switch assembly. With a high sensing temperature, the thermostatic switch is closed and the compressor clutch is energized. As the evaporator outlet temperature drops to a preset level, the thermostatic switch opens the circuit to the compressor clutch. The compressor then ceases to operate until such time as the evaporator temperature rises above the switch setting. The term *cycling clutch* derives from this on-and-off operation. In effect, the thermostatic switch is calibrated to allow the lowest possible evaporator outlet temperature that prevents the freezing of condensation on the evaporator.

Variations of the cycling clutch system include a system with a TEV and a system with an orifice tube.

Compressor Controls

There are many controls used to monitor and trigger the compressor during its operational cycle. Each of these represents the most common protective control devices designed to ensure safe and reliable operation of the compressor.

An ambient temperature switch senses outside air temperature and is designed to prevent compressor clutch engagement when air conditioning is not required or when compressor operation might cause internal damage to seals and other parts.

On some vehicles, the ambient switch is located in the air inlet duct of the air conditioning system or near the radiator. It is not required on systems with a thermostatic or pressure switch.

In cycling clutch systems, a thermostatic switch is placed in series with the compressor clutch circuit so it can turn the clutch on or off. It deenergizes the clutch and stops the compressor if the evaporator is at the freezing point. It also may control the air temperature by turning the compressor on and off intermittently.

When the temperature of the evaporator approaches the freezing point, the thermostatic switch opens the circuit and disengages the compressor clutch. The compressor remains inoperative until the evaporator temperature rises to the preset temperature, at which time the switch closes and compressor operation resumes.

A pressure cycling switch is connected in series with the compressor clutch. Like the thermostatic switch, the turning on and off of the pressure cycling switch controls the operation of the compressor.

The low-pressure cutoff or discharge pressure switch is located on the high side of the system and senses any low-pressure conditions. This switch is tied into the compressor clutch circuit, allowing it to immediately disengage the clutch when the pressure falls too low.

The high-pressure cutoff switch is normally located near the compressor and is also wired in series with the compressor clutch. This switch is designed to open and disengage the clutch at a specified high pressure. When the pressure drops below a specified pressure, it again closes and reengages the clutch.

A high-pressure relief valve may be installed on the receiver/drier, compressor, or elsewhere in the high side of the system. It is a high-pressure protection device that opens to bleed off any excessive pressure in the system.

The compressor control valve regulates the crankcase pressure in some compressors. It has a pressure-sensitive bellows assembly exposed to the suction side. This bellows acts on a ball-and-pin valve, which is exposed to high-side pressure. The bellows also controls a bleed port that is exposed to the low side. The control valve is continuously modulating, changing the displacement of the compressor according to pressure or temperature.

An electronic cycling clutch switch prevents evaporator freeze-up by sending a signal to the Powertrain Control Module (PCM). The PCM, in turn, cycles the compressor on and off by monitoring suction line temperature. If the temperature gets too low, the switch opens the input circuit to the computer, which causes the A/C clutch relay to open, which disengages the compressor clutch. Often this switch is the thermostatic switch at the evaporator.

Case/Duct Systems

The purpose of a heater/air conditioner/case and duct system is twofold. It is used to house the heater core and the evaporator and to direct the selected supply air through these components into the passenger compartment. The selected supply air can be either fresh (outside) or recirculated air, depending on the system mode. After the air is heated or cooled, it is delivered to the floor outlet, dash panel outlets, or the defrost outlets.

Manufacturers use two basic types of duct systems. The stacked core reheat system uses the heater control valve as the main control. All air enters the passenger compartment through the heater core. An access door controls only fresh or recirculated air. The blend air reheat system uses a mixture of heated and cooled air to provide comfort. The duct system brings in outside air and/or recirculated air across the heater core and evaporator. The door controls the source of the air entering into the passenger compartment.

Temperature Control Systems

Temperature control systems for air conditioners usually are connected with heater controls. Most heater and air conditioning systems use the same plenum chamber for air distribution. Two types of air conditioning controls are used: manual/semiautomatic and automatic.

Air conditioner manual/semiautomatic temperature controls operate in a manner similar to heater controls. Depending on the control setting, doors are opened and closed to direct airflow. The amount of cooling is controlled manually through the use of control settings and blower speed.

Automatic or electronic temperature control systems maintain a specific temperature automatically. To maintain a selected temperature, heat sensors send signals to a computer unit that controls compressor, heater valve, blower, and plenum door operation. A typical electronic control system might contain a coolant temperature sensor, in-car temperature sensor, outside temperature sensor, high-side temperature switch, low-side temperature switch, low-pressure switch, vehicle speed sensor, throttle position sensor, sunload sensor, and power steering cutout switch.

Three types of control panels may be found: manual, push-button, or touch pad. All serve the same purpose—they provide operator input control for the air conditioning and heating system. Some control panels have features that other panels do not have, such as provisions to display in-car and outside air temperature in degrees. Provisions are made on the control panel for operator selection of an in-car temperature between 65° and 85°F in one-degree increments. Some have an override feature that provides for a setting of either 60° or 90°F. Either of these two settings overrides all in-car temperature control circuits to provide maximum cooling or heating conditions.

SAFETY

In an automotive repair shop, there is great potential for serious accidents simply because of the nature of the business and the equipment used. Through carelessness, the automotive repair industry can be one of the most dangerous occupations. However, the chances of being injured while working on a car are close to nil if you learn to work safely and use common sense. Safety is the responsibility of everyone in the shop.

Personal Protection

Some procedures, such as grinding, result in tiny particles of metal and dust that are thrown off at very high speeds. These metal and dirt particles can easily get into your eyes, causing scratches or cuts on your eyeball. Pressurized gases and liquids escaping a ruptured or damaged hose fitting can spray a great distance. If these chemicals get into your eyes, they can cause blindness. This is especially true of refrigerants. Refrigerants can quickly freeze an eye and cause serious damage.

Eye protection should be worn whenever you are exposed to these risks. To be safe, you should wear safety glasses whenever you are working in the shop. Some procedures may require that you wear other eye protection in addition to safety glasses; for example, when cleaning parts with a pressurized spray, you should wear a face shield. The face shield not only gives added protection to your eyes but it also protects the rest of your face.

If chemicals such as battery acid, fuel, or solvents get into your eyes, flush them continuously with clean water. Have someone call a doctor and get medical help immediately.

Your clothing should be well fitted and comfortable but made with strong material. Loose, baggy clothing can easily get caught in moving parts and machinery. Some technicians prefer to wear coveralls or shop coats to protect their personal clothing. Your work clothing should offer you some protection but should not restrict your movement.

Long hair and loose, hanging jewelry can create the same type of hazard as loose-fitting clothing. They can get caught in moving engine parts and machinery. If you have long hair, tie it back or tuck it under a cap.

Never wear rings, watches, bracelets, or neck chains. These can easily get caught in moving parts and cause serious injury.

Always wear leather or similar material shoes or boots with non-slip soles. Steel-tipped safety shoes can give added protection to your feet. Jogging or basketball shoes, street shoes, or sandals are inappropriate in the shop.

Good hand protection is often overlooked. A scrape, cut, or burn can limit your effectiveness at work for many days. A well-fitted pair of heavy work gloves should be worn during operations such as grinding and welding or when handling high-temperature components. Always wear approved rubber gloves when handling strong and dangerous caustic chemicals.

When working on A/C systems, protect your hands from exposure to the refrigerant. Escaping refrigerant can cause frostbite to the area of skin exposed to the leak. In addition to wearing gloves, it is wise to use a rag when connecting and disconnecting service gauges to an A/C system.

Many technicians wear thin, surgical-type latex gloves whenever they are working on vehicles. These offer little protection against cuts but do offer protection against disease and grease buildup under and around your fingernails. These gloves are comfortable and are quite inexpensive.

Accidents can be prevented simply by the way you act. Following are some guidelines to follow while working in a shop. The list does not include everything you should or shouldn't do; it merely gives some things to think about.

- Never smoke while working on a vehicle or while working with any machine in the shop.
- Playing around is not fun when it sends someone to the hospital.
- To prevent serious burns, keep your skin away from hot metal parts, such as the radiator, exhaust manifold, tailpipe, catalytic converter, and muffler.

- Always disconnect electric engine cooling fans when working around the radiator. Many of these will turn on without warning and can easily chop off a finger or hand. Make sure you reconnect the fan after you have completed your repairs.
- When working with a hydraulic press, make sure the pressure is applied in a safe manner. It is generally wise to stand to the side when operating the press.
- Properly store all parts and tools by putting them away in a place where people will not trip over them. This practice not only cuts down on injuries, it also reduces time wasted looking for a misplaced part or tool.

Work Area Safety

Your entire work area should be kept clean and safe. Any oil, coolant, or grease on the floor can make it slippery. To clean up oil, use commercial oil absorbent. Keep all water off the floor. Water is slippery on smooth floors, and electricity flows well through water. Aisles and walkways should be kept clean and wide enough to move through easily. Make sure the work areas around machines are large enough to safely operate the machine.

Gasoline is a highly flammable volatile liquid. Something that is *flammable* catches fire and burns easily. A *volatile* liquid is one that vaporizes very quickly. *Flammable volatile* liquids are potential firebombs. Always keep gasoline or diesel fuel in an approved safety can, and never use gasoline to clean your hands or tools.

Handle all solvents (or any liquids) with care to avoid spillage. Keep all solvent containers closed, except when pouring. Proper ventilation is very important in areas where volatile solvents and chemicals are used. Solvent and other combustible materials must be stored in approved and designated storage cabinets or rooms with adequate ventilation. Never light matches or smoke near flammable solvents and chemicals, including battery acids.

Oily rags should also be stored in an approved metal container. When oily, greasy, or paint-soaked rags are left lying about or are not stored properly, they can spontaneously combust. Spontaneous combustion refers to fire that starts by itself, without a match.

Disconnecting the vehicle's battery before working on the electrical system, or before welding, can prevent fires caused by a vehicle's electrical system. To disconnect the battery, remove the negative or ground cable from the battery and position it away from the battery.

Know where all the shop's fire extinguishers are located. Fire extinguishers are clearly labeled as to type and types of fire they should be used on. Make sure you use the correct type of extinguisher for the type of fire you are dealing with. A multipurpose dry chemical fire extinguisher will put out ordinary combustibles, flammable liquids, and electrical fires. Never put water on a gasoline fire—the water will just spread the fire. The proper fire extinguisher smothers the flames.

During a fire, never open doors or windows unless it is absolutely necessary; the extra draft will only make the fire worse. Make sure the fire department is contacted before or during your attempt to extinguish a fire.

Tool and Equipment Safety

Careless use of simple hand tools, such as wrenches, screwdrivers, and hammers, causes many shop accidents that could be prevented. Keep all hand tools free of grease and in good condition. Tools that slip can cause cuts and bruises. If a tool slips and falls into a moving part, it can fly out and cause serious injury.

Use the proper tool for the job, and make sure the tool is of professional quality. Using poorly made tools or the wrong tools can damage parts, the tool itself, or you. Never use broken or damaged tools.

Safety around power tools is very important. Serious injury can result from carelessness. Always wear safety glasses when using power tools. If the tool is electrically powered, make sure it is properly grounded. Before using it, check the wiring for bare wires and for cracks in the insulation. When using electrical power tools, never stand on a wet or damp floor. Never leave a running power tool unattended.

When using compressed air, wear safety glasses or a face shield, or both. Particles of dirt and pieces of metal, blown by the high-pressure air, can penetrate your skin or get into your eyes.

Before using a compressed air tool, check all hose connections. Always hold an air nozzle or air control device securely when starting or shutting off the compressed air. A loose nozzle can whip suddenly and cause serious injury. Never point an air nozzle at anyone. Never use compressed air to blow dirt from your clothes or hair. Never use compressed air to clean the floor or workbench.

Always be careful when raising a vehicle on a lift or a hoist. Adapters and hoist plates must be positioned correctly to prevent damage to the vehicle's underbody. There are specific lift points that allow the weight of the vehicle to be evenly supported by the adapters or hoist plates. The correct lift points can be found in the vehicle's service manual. Before operating any lift or hoist, carefully read the operating manual and follow the operating instructions.

Once you know the lift supports are properly positioned under the vehicle, raise the lift until the supports contact the vehicle. Then, check the supports to make sure they are in full contact with the vehicle. Shake the vehicle to make sure it is securely balanced on the lift, and then raise the lift to the desired working height. Before working under a car, make sure the lift's locking devices are engaged.

A vehicle can be raised off the ground by a hydraulic jack. The jack's lifting pad must be positioned under an area of the vehicle's frame or at one of the manufacturer's recommended lift points. Never place the pad under the floor pan or under steering and suspension components, which are easily damaged by the weight of the vehicle. Always position the jack so the wheels of the vehicle can roll as the vehicle is being raised.

Safety stands, also called jack stands, should be placed under a sturdy chassis member, such as the frame or axle housing, to support the vehicle after it has raised by a jack. Once the safety stands are in position, the hydraulic pressure in the jack should be slowly released until the weight of the vehicle is on the stands. Never move under a vehicle when it is supported only by a hydraulic jack. Rest the vehicle on the safety stands before moving under the vehicle.

Cleaning parts is a necessary step in most repair procedures. Always wear the appropriate protection when using chemical, abrasive, and thermal cleaners.

Vehicle Operation

When the customer brings a vehicle in for service, certain driving rules should be followed to ensure your safety and the safety of those working around you. For example, before moving a car into the shop, buckle your safety belt. Make sure no one is near, the way is clear, and there are no tools or parts under the car before you start the engine.

Check the brakes before putting the vehicle in gear. Then, drive slowly and carefully in and around the shop.

If the engine must be running while work is done on the car, block the wheels to prevent the car from moving. Place the transmission in park for automatic transmissions or in neutral for manual transmissions. Set the parking (emergency) brake. Never stand directly in front of or behind a running vehicle.

Run the engine only in a well-ventilated area to avoid the danger of poisonous carbon monoxide (CO) in the engine exhaust. CO is an odorless but deadly gas. Most shops have an exhaust ventilation system; always use it. Connect the hose from the vehicle's tailpipe to the intake for the vent system. Make sure the vent system is turned on before running the engine. If the work area does not have an exhaust venting system, use a hose to direct the exhaust out of the building.

HAZARDOUS MATERIALS AND WASTES

A typical shop contains many potential health hazards for those working in it. These hazards can cause injury, sickness, impairment, discomfort, and even death. Here is a short list of the different classes of hazards.

- Chemical hazards are caused by high concentrations of vapors, gases, or solids in the form of dust.
- Hazardous wastes are substances that are the result of a service.
- Physical hazards include excessive noise, vibration, pressures, and temperatures.
- Ergonomic hazards are conditions that impede normal or proper body position and motion.

There are many government agencies charged with ensuring safe work environments for all workers. These include the Occupational Safety and Health Administration (OSHA), Mine Safety and Health Administration (MSHA), and National Institute for Occupational Safety and Health

(NIOSH). These agencies, in addition to state and local governments, have instituted regulations that must be understood and followed. Everyone in a shop is responsible for adhering to these regulations.

An important part of a safe work environment is the employees' knowledge of potential hazards. Right-to-know laws concerning all chemicals protect every employee in the shop. The general intent of right-to-know laws is for employers to provide their employees with a safe working place as it relates to hazardous materials.

All employees must be trained about their rights under the legislation, the nature of the hazardous chemicals in their workplace, and the contents of the labels on the chemicals. All information about each chemical must be posted on Material Safety Data Sheets (MSDS) and must be accessible. The manufacturer of the chemical must give these sheets to its customers if it is asked to do so. They detail the chemical composition and precautionary information for all products that can present a health or safety hazard.

Employees must become familiar with the general uses, protective equipment, accident or spill procedures, and any other information regarding the safe handling of the hazardous material. This training must be given to employees annually and provided to new employees as part of their job orientation.

All hazardous material must be properly labeled, indicating what health, fire, or reactivity hazard it poses and what protective equipment is necessary when handling each chemical. The manufacturer of the hazardous materials must provide all warnings and precautionary information, which must be read and understood by the user before use. A list of all hazardous materials used in the shop must be posted for the employees to see.

Shops must maintain documentation on the hazardous chemicals in the workplace, proof of training programs, records of accidents or spill incidents, and satisfaction of employee requests for specific chemical information via the MSDS. A general right-to-know compliance procedure manual must be used in the shop.

When handling any hazardous materials or hazardous waste, make sure you follow the required procedures for handling such material. Wear the proper safety equipment listed on the MSDS, which includes the use of approved respirator equipment.

Some of the common hazardous materials that automotive technicians use are cleaning chemicals, fuels (gasoline and diesel), paints and thinners, battery electrolyte (acid), used engine oil, refrigerants, and engine coolant (antifreeze).

Many repair and service procedures generate what are known as hazardous wastes. Dirty solvents and cleaners are good examples of hazardous wastes. Something is classified as a hazardous waste if it is on the Environmental Protection Agency (EPA) list of known harmful materials or has one or more of the following characteristics.

- *Ignitability*. A liquid with a flash point below 140°F or a solid that can spontaneously ignite.
- *Corrosivity*. A substance dissolves metals and other materials or burns the skin.
- *Reactivity*. Any material that reacts violently with water or other materials or releases cyanide gas, hydrogen sulfide gas, or similar gases when exposed to low-pH acid solutions. This includes materials that generate toxic mists, fumes, vapors, or flammable gases.
- *Toxicity*. Materials that leach one or more of eight heavy metals in concentrations greater than 100 times primary drinking water standard concentrations.

Complete EPA lists of hazardous wastes can be found in the Code of Federal Regulations. It should be noted that no material is considered hazardous waste until the shop is finished using it and ready to dispose of it.

The following list describes the recommended procedure for dealing with some of the common hazardous wastes. Always follow these and any other mandated procedures.

Refrigerants Recover or recycle refrigerants (or do both) during the service and disposal of motor vehicle air conditioner and refrigeration equipment. It is not allowable to knowingly vent refrigerants to the atmosphere. Recovering or recycling during servicing must be performed by an EPA-certified technician using certified equipment and following specified procedures.

Containers Cap, label, cover, and properly store above-ground and outdoors any liquid containers and small tanks within a diked area and on a paved impermeable surface to prevent spills from running into surface or ground water.

Oil Recycle oil. Set up equipment, such as a drip table or screen table with a used oil collection bucket, to collect oils dripping off parts. Place drip pans underneath vehicles that are leaking fluids onto the storage area. Do not mix other wastes with used oil, except as allowed by your recycler. Used oil generated by a shop (or oil received from household "do-it-yourself" generators) may be burned on site in a commercial space heater. Used oil also may be burned for energy recovery. Contact state and local authorities to determine requirements and to obtain necessary permits.

Oil filters Drain for at least 24 hours, crush, and recycle used oil filters.

Batteries Recycle batteries by sending them to a reclaimer or back to the distributor. Keeping shipping receipts can demonstrate that you have recycled. Store batteries in a watertight, acid-resistant container. Inspect batteries for cracks and leaks when they come in. Treat a dropped battery as if it were cracked. Acid residue is hazardous because it is corrosive and may contain lead and other toxics. Neutralize spilled acid by using baking soda or lime, and dispose of it as hazardous material.

Metal residue from machining Collect metal filings when machining metal parts. Keep separate and recycle if possible. Prevent metal filings from falling into a storm sewer drain.

Solvents Replace hazardous chemicals with less toxic alternatives that have equal performance. For example, substitute water-based cleaning solvents for petroleum-based solvent degreasers. To reduce the amount of solvent used when cleaning parts, use a two-stage process (dirty solvent followed by fresh solvent). Hire a hazardous waste management service to clean and recycle solvents. (Some spent solvents must be disposed of as hazardous waste unless recycled properly). Store solvents in closed containers to prevent evaporation. Evaporation of solvents contributes to ozone depletion and smog formation. In addition, the residue from evaporation must be treated as a hazardous waste. Properly label spent solvents and store in drip pans or in diked areas and only with compatible materials.

Other solids Store materials such as scrap metal, old machine parts, and worn tires under a roof or tarpaulin to protect them from the elements and to prevent potentially contaminated runoff. Consider recycling tires by retreading them.

Liquid recycling Collect and recycle coolants from radiators. Store transmission fluids, brake fluids, and solvents containing chlorinated hydrocarbons separately, and recycle or dispose of them properly.

Shop towels and rags Keep waste towels in a closed container marked "Contaminated shop towels only." To reduce costs and liabilities associated with disposal of used towels, which can be classified as hazardous wastes, investigate using a laundry service that is able to treat the wastewater generated from cleaning the towels.

Waste storage Always keep hazardous waste separate, properly labeled, and sealed in the recommended containers. The storage area should be covered and may need to be fenced and locked if vandalism could be a problem. Select a licensed hazardous waste hauler after seeking recommendations and reviewing the firm's permits and authorizations.

NATEF TASK LIST FOR HEATING AND AIR CONDITIONING

The priority rating indicates the minimum percentage of tasks, by area, a program must include in its curriculum in order to be certified in that area. Priority rating 1 has the most importance.

A. A/C System Diagnosis and Repair

A.1. Complete work order to include customer information, vehicle identifying information, customer concern, related service history, cause, and correction. Priority Rating 1.

A.2. Identify and interpret heating and air conditioning concerns; determine necessary action. Priority Rating 1.

A.3. Research applicable vehicle and service information, such as heating and air conditioning operation, vehicle service history, service precautions, and technical service bulletins. Priority Rating 1.
A.4. Locate and interpret vehicle and major component identification numbers (VIN, vehicle certification labels, calibration labels). Priority Rating 1.
A.5. Performance test A/C system; diagnose A/C system malfunctions using principles of refrigeration. Priority Rating 1.
A.6. Diagnose abnormal operating noises in the A/C system; determine necessary action. Priority Rating 2.
A.7. Identify refrigerant type; select and connect proper gauge set; record pressure readings. Priority Rating 1.
A.8. Leak test A/C system; determine necessary action. Priority Rating 1.
A.9. Inspect the condition of discharged oil; determine necessary action. Priority Rating 2.
A.10. Determine recommended oil for system application. Priority Rating 1.
A.11. Using scan tool, observe and record related HVAC data and trouble codes. Priority Rating 1.

B. Refrigeration System Component Diagnosis and Repair

1. *Compressor and Clutch*

B.1.1. Diagnose A/C system conditions that cause the protection devices (pressure, thermal, and PCM) to interrupt system operation; determine necessary action. Priority Rating 2.
B.1.2. Inspect and replace A/C compressor drive belts; determine necessary action. Priority Rating 1.
B.1.3. Inspect, test, and/or replace A/C compressor clutch components and/or assembly. Priority Rating 2.
B.1.4. Remove, inspect, and reinstall A/C compressor and mountings; determine required oil quantity. Priority Rating 1.
B.1.5. Identify hybrid vehicle AC system electrical circuits, service and safety precautions. Priority Rating 3.

2. *Evaporator, Condenser, and Related Components*

B.2.1. Determine need for an additional A/C system filter; perform necessary action. Priority Rating 1.
B.2.2. Remove and inspect AC system mufflers, hoses, lines, fittings, O-rings, seals, and service valves; perform necessary action. Priority Rating 2.
B.2.3. Inspect A/C condenser for airflow restrictions; perform necessary action. Priority Rating 1.
B.2.4. Remove, inspect, and reinstall receiver/drier or accumulator/drier; determine required oil quantity. Priority Rating 1.
B.2.5. Remove and install expansion valve or orifice (expansion) tube. Priority Rating 2.
B.2.6. Inspect evaporator housing water drain; perform necessary action. Priority Rating 3.
B.2.7. Remove, inspect, and reinstall evaporator; determine required oil quantity. Priority Rating 3.
B.2.8. Remove, inspect, and reinstall condenser; determine required oil quantity. Priority Rating 3.

C. Heating, Ventilation, and Engine Cooling Systems Diagnosis and Repair

C.1. Diagnose temperature control problems in the heater/ventilation system; determine necessary action. Priority Rating 2.
C.2. Perform cooling system pressure tests; check coolant condition, inspect and test radiator, pressure cap, coolant recovery tank, and hoses; perform necessary action. Priority Rating 1.
C.3. Inspect engine cooling and heater system hoses and belts; perform necessary action. Priority Rating 1.
C.4. Inspect, test, and replace thermostat and gasket. Priority Rating 1.
C.5. Determine coolant condition and coolant type for vehicle application; drain and recover coolant. Priority Rating 1.
C.6. Flush system; refill system with recommended coolant; bleed system. Priority Rating 1.

C.7. Inspect and test cooling fan, fan clutch, fan shroud, and air dams; perform necessary action. Priority Rating 1.
C.8. Inspect and test electrical cooling fan, fan control system, and circuits; determine necessary action. Priority Rating 1.
C.9. Inspect and test heater control valve(s); perform necessary action. Priority Rating 2.
C.10. Remove, inspect, and reinstall heater core. Priority Rating 1.

D. Operating Systems and Related Controls Diagnosis and Repair

D.1. Diagnose malfunctions in the electrical controls of heating, ventilation, and A/C (HVAC) systems; determine necessary action. Priority Rating 2.
D.2. Inspect and test A/C-heater blower, motors, resistors, switches, relays, wiring, and protection devices; perform necessary action. Priority Rating 1.
D.3. Test and diagnose A/C compressor clutch control systems; determine necessary action. Priority Rating 1.
D.4. Diagnose malfunctions in the vacuum, mechanical, and electrical components and controls of the heating, ventilation, and A/C (HVAC)system; determine necessary action. Priority Rating 2.
D.5. Inspect and test A/C-heater control panel assembly; determine necessary action. Priority Rating 3.
D.6. Inspect and test A/C-heater control cables, motors, and linkages; perform necessary action. Priority Rating 3.
D.7. Inspect and test A/C-heater ducts, doors, hoses, cabin filters, and outlets; perform necessary action. Priority Rating 3.
D.8. Check operation of automatic and semiautomatic HVAC control systems; determine necessary action. Priority Rating 3.

E. Refrigerant Recovery, Recycling, and Handling

E.1. Perform correct use and maintenance of refrigerant handling equipment. Priority Rating 1.
E.2. Identify (by label application or use of a refrigerant identifier) and recover A/C system refrigerant. Priority Rating 1.
E.3. Recycle refrigerant. Priority Rating 1.
E.4. Label and store refrigerant. Priority Rating 1.
E.5. Test recycled refrigerant for noncondensable gases. Priority Rating 1.
E.6. Evacuate and charge A/C system. Priority Rating 1.

DEFINITION OF TERMS USED IN THE TASK LIST

To clarify the intent of these tasks, NATEF has defined some of the terms used in the task listings. For a good understanding of what the task includes, refer to this glossary while reading the task list.

add To increase fluid or pressure to the correct level or amount.

adjust To bring components to specified operational settings.

assemble (reassemble) To fit together the components of a device.

bleed To remove air from a closed system.

charge To bring to "full" state (e.g., battery or air conditioning system).

check To verify condition by performing an operational or comparative examination.

clean To rid components of extraneous matter for the purpose of reconditioning, repairing, measuring, and reassembling.

determine To establish the procedure to be used to effect the necessary repair.

determine necessary action Indicates that the diagnostic routine or routines is the primary emphasis of a task. The student is required to perform the diagnostic steps and communicate the diagnostic outcomes and corrective actions required to address the concern or problem. The training program determines the communication method (worksheet, test, verbal communication, or other means deemed appropriate) and whether the corrective procedures for these tasks are actually performed.

diagnose To locate the root cause or nature of a problem by using the specified procedure.

disassemble To separate a component's parts in preparation for cleaning, inspection, or service.

discharge To empty a storage device or system.

drain To use gravity to empty a container.

evacuate To remove air, fluid, or vapor from a closed system by the use of a vacuum pump.

fill (refill) To bring fluid level to a specified point or volume.

flush To use fluid to clean an internal system.

identify To establish the identity of a vehicle or component prior to service; to determine the nature or degree of a problem.

inspect (see *check*)

install (reinstall) To place a component in its proper position in a system.

leak test To locate the source of leaks in a component or system.

locate Determine or establish a specific spot or area.

On-board diagnostics (OBD) A diagnostic system contained in the Powertrain Control Module (PCM), which monitors computer inputs and outputs for failures. OBD II is an industry-standard, second generation OBD system that monitors emissions control systems for degradation as well as failures.

perform To accomplish a procedure in accordance with established methods and standards.

perform necessary action Indicates that the student is to perform the diagnostic routine(s) and perform the corrective action item. Where various scenarios (conditions or situations) are presented in a single task, at least one of the scenarios must be accomplished.

pressure test To use air or fluid pressure to determine the integrity, condition, or operation of a component or system.

priority ratings Indicates the minimum percentage of tasks, by area, a program must include in its curriculum in order to be certified in that area.

purge To eliminate an undesired air or fluid from a closed system.

reassemble (See *assemble*)

refill (See *fill*)

remove To disconnect and separate a component from a system.

replace To exchange an unserviceable component with a new or rebuilt component; to reinstall a component.

select To choose the correct part or setting during assembly or adjustment.

service To perform a specified procedure when called for in the owner's manual or service manual.

test To verify condition through the use of meters, gauges, or instruments.

verify To establish that a problem exists after hearing the customer's complaint and performing a preliminary diagnosis.

HEATING AND AIR CONDITIONING TOOLS AND EQUIPMENT

Many different tools and pieces of testing and measuring equipment are used to service heating and air conditioning systems. NATEF has said that an air conditioning technician must know what they are and how and when to use them. The tools and equipment listed by NATEF are covered in the following discussion, along with the tools and equipment you will use while completing the job sheets. Although you will be using common hand tools, they are not part of this discussion. You should already know what they are and how to use and care for them.

NOTE: *Air conditioning systems are extremely sensitive to moisture and dirt. Therefore, clean working conditions are extremely important. The smallest particle of foreign matter in an air conditioning system contaminates the refrigerant, causing rust, ice, or damage to the compressor. For this reason, all replacement parts are sold in vacuum-sealed containers and should not be opened until they are to be installed in the system. If, for any reason, a part has been removed from its container for any length of time, the part must be completely flushed using only a recommended flush solvent to remove any dust or moisture that might have accumulated during storage. When the system has been open for any length of time, the entire system must be purged completely. A new receiver/drier must be installed because the element of the existing unit will have become saturated and unable to remove any moisture from the system once the system is recharged.*

REFRIGERANT SAFETY PRECAUTIONS

- Always work in a well-ventilated and clean area. Refrigerant (R-12 and R-134a) is colorless and invisible as a gas. Refrigerant is heavier than oxygen and will displace oxygen in a confined area. Avoid breathing the refrigerant vapors. Exposure to refrigerant may irritate your eyes, nose, and throat.
- Refrigerant evaporates quickly when it is exposed to the atmosphere. It will freeze anything it contacts. If liquid refrigerant gets in your eyes or on your skin, it can cause frostbite. Never rub your eyes or skin if you have come in contact with refrigerant. Immediately flush the exposed areas with cool water for 15 minutes and seek medical help.
- An A/C system's high pressure can cause severe injury to your eyes or skin if a hose bursts. Always wear eye protection when working around the A/C system and refrigerant. It is also advisable to wear protective gloves and clothing.
- Never use R-134a in combination with compressed air for leak testing. Pressurized R-134a, in the presence of oxygen, may form a combustible mixture. Never introduce compressed air into R-134a containers (empty or full ones), A/C systems, or A/C service equipment.
- Be careful when handling refrigerant containers. Never drop, strike, puncture, or burn the containers. Always use Department of Transportation (DOT)-approved refrigerant containers.
- Never expose A/C system components to high temperatures—heat causes the refrigerant's pressure to increase. Never expose refrigerant to an open flame—a poisonous gas (phosgene) results. If inhaled, phosgene gas can cause severe respiratory irritation. This gas also occurs when an open-flame leak detector is used. Phosgene fumes have an acrid (bitter) smell. If the refrigerant needs to be heated during service, the bottom of the refrigerant container should be placed in warm water (less than 125°F).
- Never overfill refrigerant containers. The filling level of the container should never exceed 60 percent of the container's gross weight rating. Always store refrigerant containers in temperatures below 125°F and keep them out of direct sunlight.
- Refrigerant comes in 30- and 50-pound cylinders. Remember that these drums are under considerable pressure and should be handled following precautions. Keep the drums in an upright position. Make sure that metal caps protect valves and safety

plugs when the drums are not in use. Avoid dropping the drums. Handle them carefully. When transporting refrigerant, do not place containers in the vehicle's passenger compartment.

- R-12 should be stored and sold in white containers, whereas R-134a should be stored in light blue containers. R-12 and R-134a should never be mixed. Their oils and desiccants are not compatible. If the two refrigerants are mixed, contamination will occur and may result in A/C system failure. Separate service equipment, including recovery/recycling machines and service gauges, should be used for the different refrigerants. Always read and follow the instructions from the equipment manufacturer when servicing A/C systems.
- To prevent cross-contamination, identify whether the A/C system being worked on uses R-12 or R-134a. Check the fittings in the system; all R-134a-based systems use 1/2-inch-16 ACME threaded fittings and quick-disconnect service couplings. Underhood labels clearly stating that R-134a is used can identify most R-134a systems. Most manufacturers identify the type of refrigerant used by labeling the compressor. Also look for a label with the words, *"CAUTION—SYSTEM TO BE SERVICED BY QUALIFIED PERSONNEL."* This label or plate can be found under the hood, near a component of the system. This label also tells you what kind of refrigerant and refrigerant oil are used. Before beginning to service an A/C system, determine the type of refrigerant used.
- R-134a can also be identified by viewing the sight glass, if the system has one. The appearance of R-134a will be milky due to the mixture of refrigerant and the refrigerant oil.

Manifold Gauge Set

The manifold gauge set (Figure 2) is used when discharging, charging, evacuating, and diagnosing trouble in the system. With the new legislation on handling refrigerants, all gauge sets are required to have a valve device to close off the end of the hose so that the fitting not in use is automatically shut.

Figure 2 A manifold gauge set.

The low-pressure gauge is graduated into pounds of pressure from 1 to 120 (with cushion to 250) in 1-pound graduations, and, in the opposite direction, in inches of vacuum from 0 to 30. This is the gauge that should always be used in checking pressure on the low-pressure side of the system. The high-pressure gauge is graduated from 0 to 500 pounds pressure in 10-pound graduations. This gauge is used for checking pressure on the high-pressure side of the system.

The center manifold fitting is common to both the low and the high side and is for evacuating or adding refrigerant to the system. When this fitting is not being used, it should be capped. A test hose connected to the fitting directly under the low-side gauge is used to connect the low side of the test manifold to the low side of the system. A similar connection is found on the high side.

The gauge manifold is designed to control refrigerant flow. When the manifold test set is connected into the system, pressure is registered on both gauges at all times. During all tests, both the low- and high-side hand valves are in the closed position (turned inward until the valve is seated).

Refrigerant flows around the valve stem to the respective gauges and registers the system's low-side pressure on the low-side gauge and the system's high-side pressure on the high-side gauge. The hand valves isolate the low and high sides from the central portion of the manifold. When the gauges are first connected to the gauge fittings with the refrigeration system charged, the gauge lines should always be purged. Purging is done by cracking each valve on the gauge set to allow the pressure of the refrigerant in the refrigeration system to force the air to escape through the center gauge line. Failure to purge lines can result in air or other contaminants entering the refrigeration system.

Because R-134a is not interchangeable with R-12, separate sets of hoses, gauges, and other equipment are required to service vehicles. All equipment used to service R-134a and R-12 systems must meet Society of Automotive Engineers (SAE) standard J1991. The service hoses on the manifold gauge set must have manual or automatic back-flow valves at the service port connector ends. These valves prevent the refrigerant from being released into the atmosphere during connection and disconnection. Manifold gauge sets for R-134a can be identified by labels on the gauges; in addition or instead, the face of the gauges may be light blue.

For identification purposes, R-134a service hoses must have a black stripe along their length and be clearly labeled. The low-pressure hose is blue with a black stripe. The high-pressure hose is red with a black stripe, and the center service hose is yellow with a black stripe. Service hoses for one type of refrigerant do not easily connect to the wrong system because the fittings for an R-134a system are different from those used in an R-12 system.

Service Port Adapters

The high-side fitting on many vehicles with an R-12 system may require the use of a special adapter to connect the manifold gauge set to the service port. These adapters (Figure 3) are installed to the end of the service hose before connecting the hose to the system. The service hoses of some manifold gauge sets are not equipped with a Schrader valve-depressing pin. Therefore, when connecting this type of hose to a Schrader valve, an adapter (Figure 4) must be used. The manifold and gauge sets for R-12 and R-134a are not interchangeable, therefore there are no suitable or allowable adapters for using R-12 gauges on a R-134a system, or vice versa.

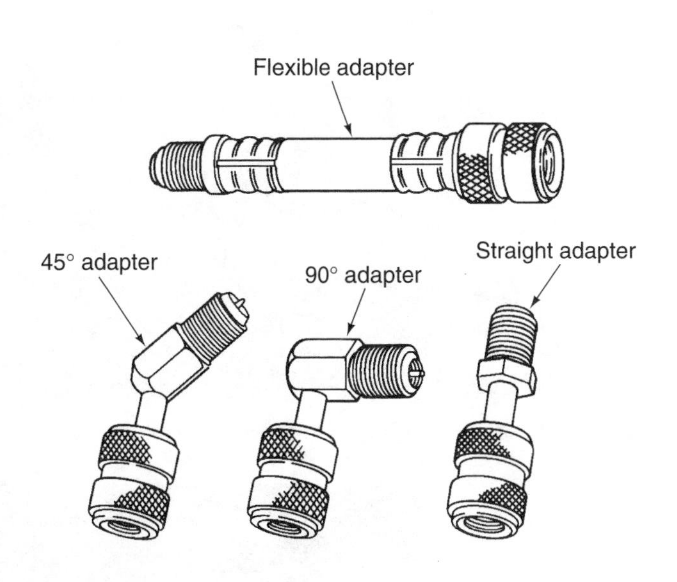

Figure 3 High-side adapters to connect the service hose to the service ports.

Electronic Leak Detector

An electronic leak detector is safe, effective, and can be used with all types of refrigerants. A handheld, battery-operated electronic leak detector contains a test probe that is moved about 1 inch per second in areas of suspected leaks (Figure 5). Since refrigerant is heavier than air, the probe should be positioned below the test point. An alarm or a buzzer on the detector indicates the presence of a leak. On some models, a light flashes when refrigerant is detected.

Fluorescent Leak Tracer

To find a refrigerant leak using the fluorescent tracer system, first introduce a fluorescent dye into the air conditioning system with a special infuser included with the detector equipment. Run the air conditioner for a few minutes, giving the tracer dye fluid time to circulate and penetrate. Wearing the tracer protective goggles, scan the system with a black-light glow gun. Leaks in

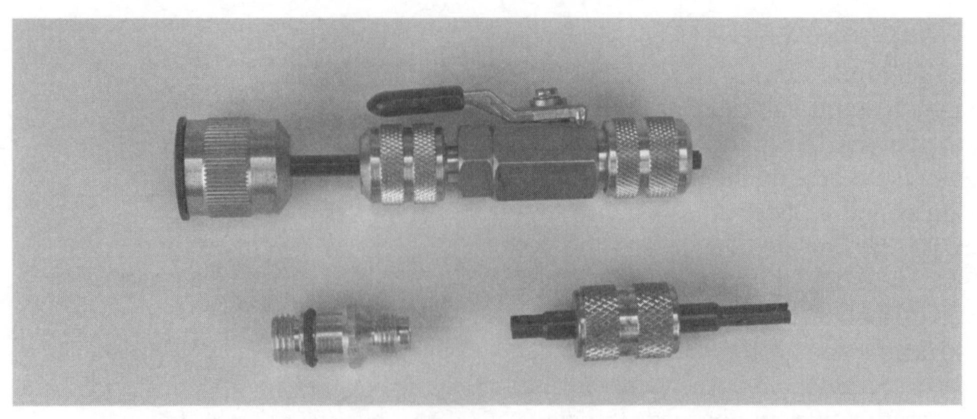

Figure 4 Schrader valve adapters for service hoses.

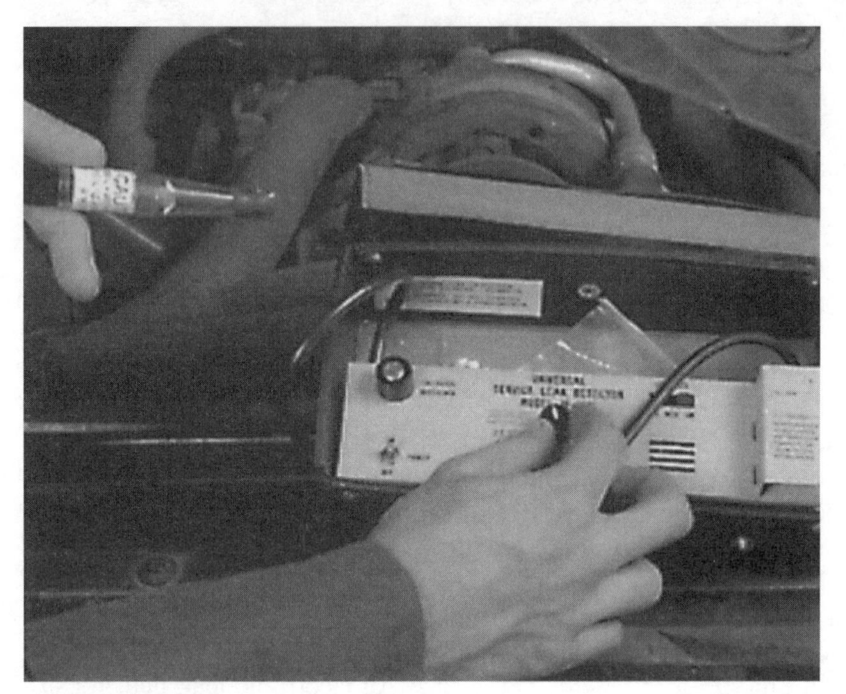

Figure 5 An electronic leak detector.

the system shine under the black light as a luminous yellow-green.

Scan Tools

The introduction of computer-controlled systems brought with it the need for tools capable of troubleshooting electronic control systems. A variety of computer scan tools are available today that do just that. A scan tool is a microprocessor designed to communicate with the vehicle's computer. Connected to the computer through diagnostic connectors, a scan tool can access trouble codes, run tests to check system operations, and monitor the activity of the system. Trouble codes and test results are displayed on an LED screen or printed out on the scanner printer.

Scan tools retrieve fault codes from a computer's memory and digitally display these codes on the tool. A scan tool may also perform many other diagnostic functions depending on the year and make of the vehicle. Most aftermarket scan tools have removable modules that are updated each year. These modules are designed to test the computer systems on various makes of vehicles. For example, some scan testers have a 3-in-1 module that tests the computer systems on Chrysler, Ford, and General Motors vehicles. A 10-in-1 module is also available to diagnose computer systems on vehicles imported by ten different manufacturers. These modules plug into the scan tool.

Scan tools are capable of testing many onboard computer systems, such as climate controls, transmission controls, engine computers, antilock brake computers, air bag computers, and suspension computers, depending on the year and make of the vehicle and the type of scan tester. In many cases, the technician must select the computer system to be tested with the scanner after it has been connected to the vehicle.

The scan tool is connected to specific diagnostic connectors on various vehicles. Most manufacturers have one diagnostic connector that connects the data wire from each onboard computer to a specific terminal in this connector. Other vehicle manufacturers have several diagnostic connectors on each vehicle, and each of these connectors may be connected to one or more onboard computers. A set of connectors is supplied with the scanner to allow tester connection to various diagnostic connectors on different vehicles.

The scanner must be programmed for the model year, make of vehicle, and type of engine. With some scan tools, this selection is made by pressing the appropriate buttons on the tester, as directed by the digital tester display. On other scan testers, the appropriate memory card must be installed in the tester for the vehicle being tested. Some scan testers have a built-in printer to print test results, whereas other scan testers may be connected to an external printer.

As automotive computer systems become more complex, the diagnostic capabilities of scan testers continue to expand. Many scan testers now have the capability to store, or *freeze,* data into the tester during a road test, and then play the data back when the vehicle is returned to the shop.

Some scan testers now display diagnostic information based on the fault code in the computer memory. Service bulletins published by the manufacturer of the scan tester may be indexed by the tester after the vehicle information is entered in the tester. Other scan testers will display sensor specifications for the vehicle being tested.

The vehicle's computer sets trouble codes when a voltage signal is entirely out of its normal range. The codes help technicians identify the cause of the problem. If a signal is within its normal range but is still not correct, the vehicle's computer does not display a trouble code, though a problem may still exist.

With the On-Board Diagnostic II system (OBD-II), the diagnostic connectors are located in the same place on all vehicles. Also, any scan tools designed for OBD-II work on all OBD-II systems; therefore, the need for designated scan tools or cartridges is eliminated. The OBD-II scan tool has the ability to run diagnostic tests on all systems and has "freeze-frame" capabilities.

Stethoscope

Some sounds can be easily heard without using a listening device, but others are impossible to hear unless amplified. A stethoscope is very helpful in locating the cause of a noise by amplifying the sound waves. It can also help you distinguish between normal and abnormal noise. The procedure for using a stethoscope is simple. Use the metal prod to trace the sound until it reaches its maximum intensity. Once the precise location has been discovered, the sound can be better evaluated. A sounding stick, which is nothing more than a long, hollow tube, works on the same principle, though a stethoscope gives much clearer results.

The best results, however, are obtained with an electronic listening device. With this tool you can tune into the noise. Doing this allows you to eliminate all other noises that might distract or mislead you.

Belt Tension Gauge

A belt tension gauge is used to measure drive belt tension. The belt tension gauge is installed over the belt, and the gauge indicates the amount of belt tension.

Coolant Hydrometer

A coolant hydrometer is used to check the amount of antifreeze in the coolant. This tester contains a pickup hose, coolant reservoir, and squeeze bulb. The pickup hose is placed in the radiator coolant. When the squeeze bulb is squeezed and released, coolant is drawn into the reservoir. As coolant enters the reservoir, a pivoted float moves upward with the coolant level. A pointer on the float indicates the freezing point of the coolant on a scale located on the reservoir housing.

Cooling System Pressure Tester

A cooling system pressure tester contains a hand pump and a pressure gauge. A hose is connected from the hand pump to a special adapter that fits on the radiator filler neck. This tester is used to pressurize the cooling system and check for coolant leaks. Additional adapters are available to connect the tester to the radiator cap. With the tester connected to the radiator cap, the pressure-relief action of the cap may be checked.

Thermometer

A dial-type thermometer (Figure 6) is often used to measure the air temperature at the vent outlets. This procedure checks the overall performance of the system. The thermometer can also be used to check the temperature of refrigerant lines, hoses, and components while diagnosing a system. An electronic pyrometer works best for this procedure and is often used.

Feeler Gauge

A feeler gauge is a thin strip of metal or plastic of known and closely controlled thickness. Several of these strips are often assembled together as a feeler gauge set, which looks like a pocketknife. The desired thickness gauge can be pivoted away from others for convenient use. A feeler gauge set usually contains strips or leaves of 0.002- to 0.010-inch thickness (in steps of 0.001 inch) and leaves of 0.012- to 0.024-inch thickness (in steps of 0.002 inch).

A feeler gauge can be used by itself to measure clearances, such as critical compressor clutch clearances. It can also be used with a precision straightedge to measure the flatness of sealing surfaces in a compressor.

Straightedge

A straightedge is no more than a flat bar machined to be totally flat and straight; to be effective, it must be flat and straight. Any surface that should be flat can be checked with a straightedge and feeler gauge set. The straightedge is placed across and at angles on the surface. At any low points on the surface, a feeler gauge can be placed between the straightedge and the surface. The size gauge that fills in the gap is the amount of warpage or distortion.

Dial Indicator

The dial indicator is calibrated in 0.001-inch (one-thousandth-inch) increments. Metric dial indicators are also available; both types are used to measure movement. Common uses of the dial indicator include checking compressor shaft run out, shaft endplay, and compressor clutch run out.

To use a dial indicator, position the indicator rod against the object to be measured. Then, push the indicator toward the work until the indicator needle travels far enough around the gauge face to permit movement to be read in either direction. Zero the indicator needle on the gauge. Move the object in the direction required while observing the needle of the gauge. Always be sure the range of the dial indicator is sufficient to allow the amount of movement required by the measuring procedure. For example, never use a 1-inch indicator on a component that will move 2 inches.

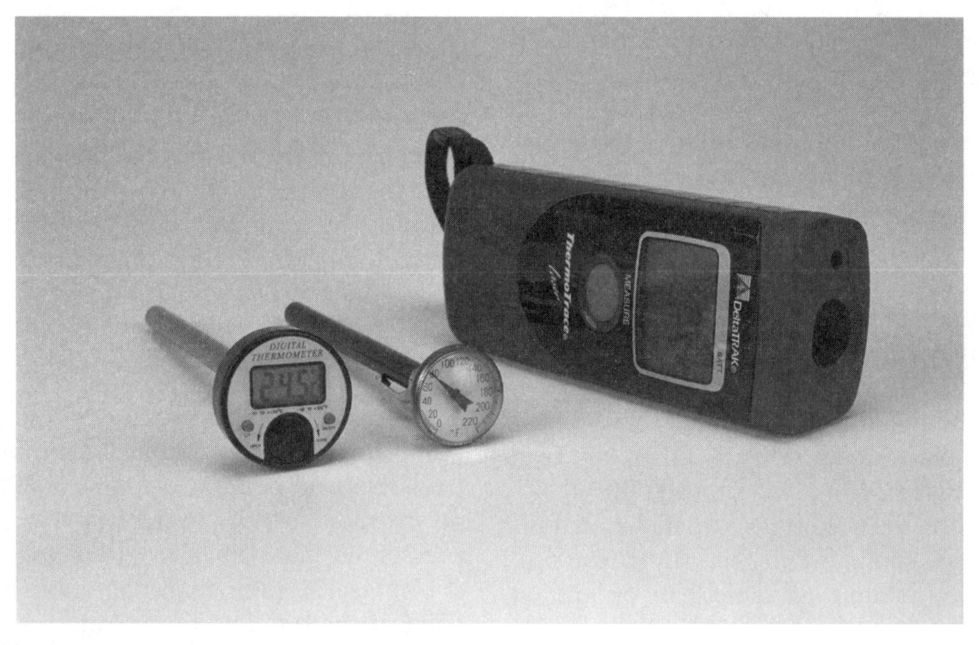

Figure 6 Three types of commonly used thermometers.

Torque-Indicating Wrench

Torque is the twisting force used to turn a fastener against the friction between the threads and between the head of the fastener and the surface of the component. The fact that practically every vehicle and engine manufacturer publishes a list of torque recommendations is ample proof of the importance of using proper amounts of torque when tightening nuts or bolts. The amount of torque applied to a fastener is measured with a torque-indicating, or torque, wrench.

Three basic types of torque-indicating wrenches are available, in pound-per-inch and pound-per-foot increments. A beam torque wrench has a beam that points to the torque reading; a click-type torque wrench has the desired torque reading set on the handle (when the torque reaches that level, the wrench clicks); and a dial torque wrench has a dial that indicates the torque exerted on the wrench. Some designs of the dial torque wrench have a light or buzzer that turns on when the desired torque is reached.

Gear and Bearing Pullers

Many tools are designed for a specific purpose. An example of a special tool is a gear and bearing puller. Many gears and bearings have a slight interference fit (press fit) when they are installed on a shaft or in a housing. Something that has a press fit has an interference fit. For example, the inside diameter of a bore is 0.001 inch smaller than the outside diameter of a shaft, and when the shaft is fitted into the bore, it must be pressed in to overcome the 0.001-inch interference fit. This press fit prevents the parts from moving on each other. The removal of these gears and bearings must be done carefully to prevent damage to the gears, bearings, or shafts. Prying or hammering can break or bind the parts. A puller with the proper jaws and adapters should be used to remove gears and bearings. Using the proper puller, the force required to remove a gear or bearing can be applied with a slight and steady motion.

Bushing and Seal Pullers and Drivers

Another commonly used group of special tools are the various designs of bushing and seal drivers and pullers. Pullers are either a threaded or slide-hammer type of tool. Always make sure you use the correct tool for the job because bushings and seals are easily damaged if the wrong tool or procedure is used. Car manufacturers and specialty tool companies work together closely to design and manufacture special tools required to repair cars. Most of these special tools are listed in the appropriate service manuals.

Retaining Ring Pliers

An air conditioning technician will encounter many different styles and sizes of retaining rings and snap rings that hold things together or keep them in a fixed location. Using the correct tool to remove and install these rings is the only safe way to work with them. All technicians should have an assortment of snap ring and retaining ring pliers.

Compressor Tools

Although compressors are typically replaced when they are faulty, certain service procedures applied to them are standard practice. Most of these procedures focus on compressor clutch and shaft seal service, and they require special tools (Figure 7). Clutch plate tools are required to gain access to the shaft seal. They are also needed to reinstall the clutch plate after service.

Many different tools are required to perform these services on a compressor. Typically, to replace a shaft seal, you need an adjustable or fixed spanner wrench, clutch plate installer/remover, ceramic seal installer/remover, seal assembly installer/remover, seal seat installer/remover, shaft seal protector, snap ring pliers, O-ring remover, and O-ring installer. Some of these tools are made for a specific model compressor; others are universal fit or will have interchangeable parts to allow them to work on a variety of compressors.

Can Tap

A can tap is used to connect a 1-"pound" can of refrigerant to the manifold gauge set. These cans are available in either screw top or flat top. The can tap either pierces through the top or makes a screw-type connection with the top. It should be noted that "1-pound" cans actually contain only 12 or 14 ounces. Many states have restrictions or strict prohibitions on the sale and use of small 1-pound cans of refrigerant.

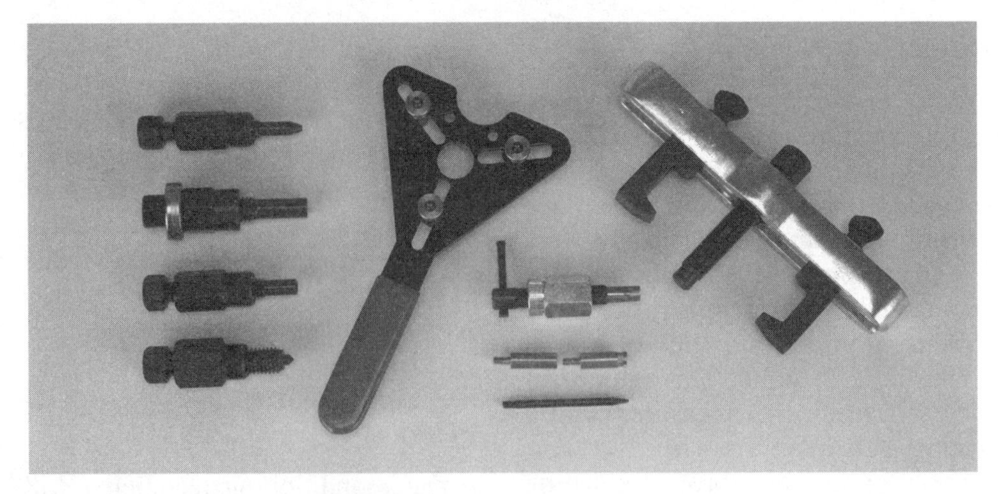

Figure 7 A sample of common compressor tools.

Hose and Fitting Tools

An A/C system is a closed system, meaning outside air should never enter the system, and the refrigerant in the system should never exit to the outside. To maintain its integrity, special fittings and hoses are used. Often, special tools are required when servicing the system's fittings and hoses. An example of this is the spring-lock coupling tool set; without the tool, it is impossible to separate this connector without damaging it.

Charging Cylinder

The charging cylinder is designed to mete out a desired amount of refrigerant by weight. Compensation for temperature variations is accomplished by reading the pressure on the gauge of the cylinder and dialing the plastic shroud. The calibrated chart on the shroud contains corresponding pressure readings for the refrigerant being used.

Refrigerant Identifier

A refrigerant identifier (Figure 8), obviously, is used to identify the type of refrigerant that is present in a system. This needs to be done before any service work begins. The tester is used to identify the purity and quality of the refrigerant sample taken from the system.

Refrigerant Recovery/Recycling System

Each shop that does A/C work must have a recovery, recycle, and recharge machine (Figure 9) for each type of refrigerant. There are some recovery, recycle, and recharge machines that have the capability of working with R-12 and R-134a, but check the machine before attempting to do so.

There are currently two types of refrigerant recovery machines, the single-pass and the multipass. Both have the ability to draw the refrigerant from the vehicle, filter and separate the oil from it, remove moisture and air from it, and store the refrigerant until it is reused.

In a single-pass system, the refrigerant goes through each stage before being stored. In multipass systems, the refrigerant may go through all stages or some of the stages before being stored. Either system is acceptable if it has the UL-approved label.

All recycled refrigerant must be safely stored in DOT CFR Title 49-approved or UL-approved containers. Containers specifically made for R-134a should be so marked. Before any container of recycled refrigerant can be used, it must be checked for noncondensable gases.

Vacuum Pump

It is critical that all moisture be removed from a system before it is recharged. Moisture is removed by connecting a vacuum pump to the system. The vacuum pump may be a separate item or may be incorporated into other air conditioning equipment, such as the recovery/recycling machine.

The main purpose of the vacuum pump is to remove the contaminating air and moisture from the system. The vacuum pump reduces system pressure in order to vaporize the moisture and then exhausts the vapor along with all remaining air. The pump's ability to clean the system is directly related to its ability to reduce pressure (create a vacuum) low enough to boil off all the contaminating moisture.

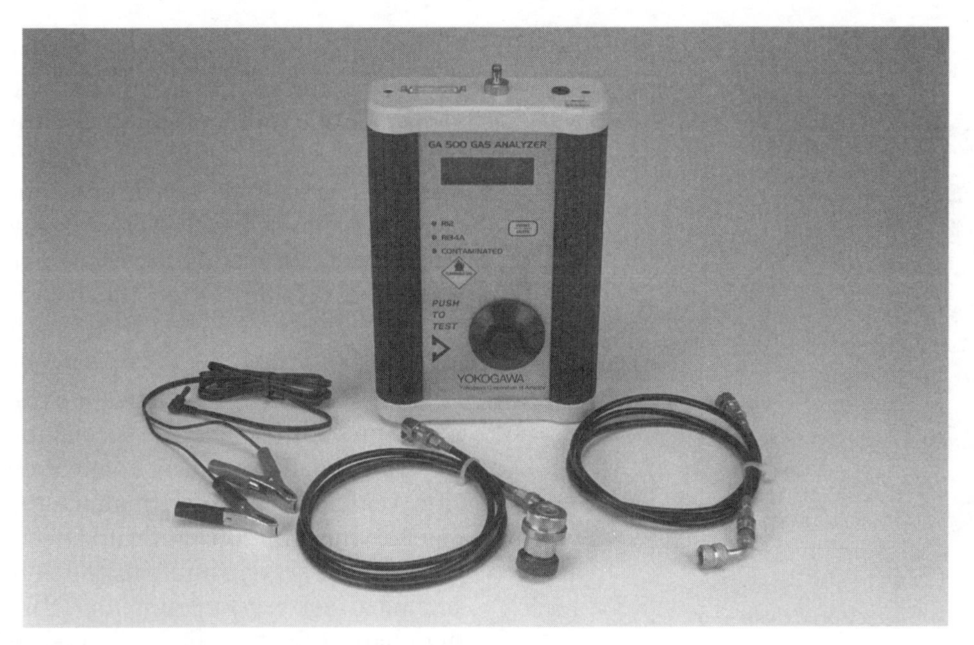

Figure 8 A refrigerant identifier.

Thermistor Vacuum Gauge

An electronic thermistor vacuum gauge is designed to work with the vacuum pump to measure the last, most critical inch of mercury during evacuation. It constantly monitors and visually indicates the vacuum level so that you know when the system has a full vacuum and will be free of moisture.

After the system is evacuated, it can be recharged. If the system will not pull down to a good vacuum, there is probably a leak somewhere in the system.

Electronic Scale

An electronic scale (Figure 10) is used to ensure that an accurate amount of refrigerant has been installed in a system. Since refrigerant is installed by weight, precise measurements of the refrigerant container before and after charging the system indicates how much refrigerant has been added to the system.

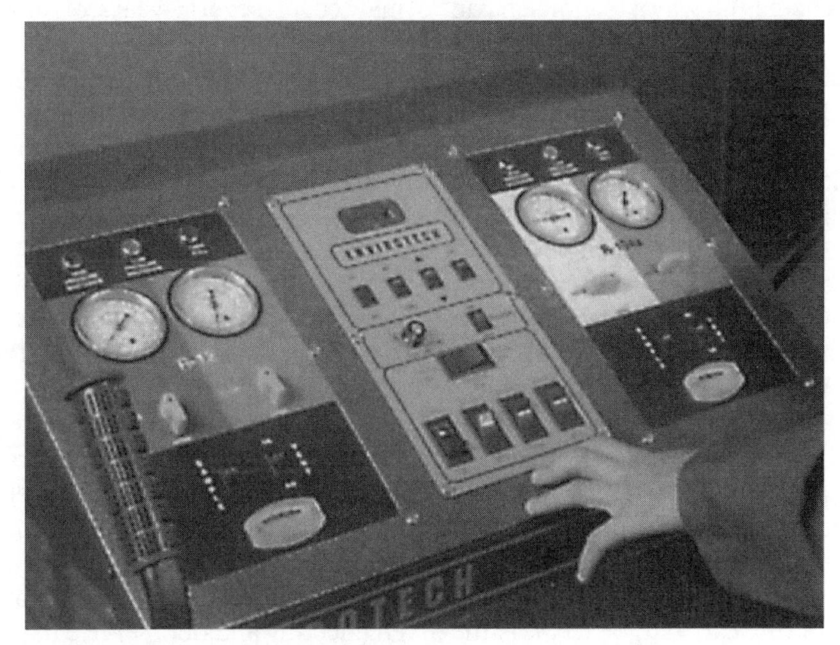

Figure 9 A refrigerant recovery/recycling machine.

Figure 10 An electronic scale.

Coolant Recovery and Recycle System

A coolant recovery and recycle machine typically can drain, recycle, fill, flush, and pressure test a cooling system. Typically, additives are mixed into the used coolant during recycling. These additives either bind to contaminants in the coolant so they can be easily removed, or they restore some of the chemical properties in the coolant.

Service Manuals

Perhaps the most important tools you will use are service manuals. There is no way a technician can remember all the procedures and specifications needed to correctly repair all vehicles. Thus, a good technician relies on service manuals and other information sources for this information. Good information plus knowledge allows a technician to fix a problem with the least amount of frustration and at the lowest cost to the customer.

To obtain the correct system specifications and other information, you must first identify the exact system you are working on. The best source for vehicle identification is the vehicle identification number (VIN). The code can be interpreted through information given in the service manual. In addition, the manual may also help you recognize the system through identification of key components or other numbers or markings.

The primary source of repair and specification information for any car, van, or truck is the manufacturer. The manufacturer publishes service manuals each year for every vehicle built. Because of the enormous amount of information, some manufacturers publish more than one manual per year per car model. They are typically divided into sections based on the major systems of the vehicle. In the case of air conditioning systems, there is a section for each engine that may be in the vehicle. Often the heating system is included with information about the engine's cooling system. Manufacturers' manuals cover all repairs, adjustments, specifications, detailed diagnostic procedures, and the special tools required.

Since many technical changes occur on specific vehicles each year, manufacturers' service manuals must be constantly updated. Updates are published as service bulletins (often referred to as Technical Service Bulletins, or TSBs) that show the changes in specifications and repair procedures during the model year. These changes do not appear in the service manual until the next year. The car manufacturer provides these bulletins to dealers and repair facilities on a regular basis.

Service manuals are also published by independent companies rather than the manufacturers. However, they pay for and get most of their information from the car makers. They contain component information, diagnostic steps, repair procedures, and specifications for several car makes in one book. Information is usually condensed and is more general in nature than that in the manufacturer's manuals. The condensed format allows for more coverage in less space and, therefore, is not always specific. One book may cover several years of models and several car makes.

Many of the larger parts manufacturers have excellent guides on the various parts they manufacture or supply. They also provide updated service bulletins on their products. Other sources for up-to-date technical information are trade magazines and trade associations.

The same information that is available in service manuals is now commonly found electronically on compact disks (CD-ROMs), digital video disks (DVDs), and the Internet. A single compact disk can hold a quarter million pages of text, eliminating the need for a huge library to contain all of the printed manuals. Using electronics to find information is also easier and quicker. The disks are normally updated quarterly and not only contain the most recent service bulletins but also engineering and field service fixes. DVDs can hold more information than CDs; therefore, fewer disks are needed with systems that use DVDs. The CDs and DVDs are inserted into a computer. All a technician needs to do is enter vehicle information and then move to the appropriate part or system. The appropriate information will then appear on the computer's screen. Online data can be updated instantly and requires no space for physical storage. These systems are easy to use and the information is quickly accessed and displayed. The computer's keyword, mouse, and/or light pen are used to make selections from the screen's menu. Once the information is retrieved, a technician can read it off the screen or print it out and take it to the service bay.

CROSS-REFERENCE GUIDE

NATEF Task	Job Sheet
A.1	1
A.2	2
A.3	3
A.4	3
A.5	4
A.6	5
A.7	6 & 7
A.8	8
A.9	9
A.10	9
A.11	10
B.1.1	11
B.1.2	12
B.1.3	11
B.1.4	13
B.1.5	14
B.2.1	15
B.2.2	16
B.2.3	5
B.2.4	16
B.2.5	16
B.2.6	17
B.2.7	18
B.2.8	19
C.1	20
C.2	21
C.3	22
C.4	23
C.5	21
C.6	21
C.7	24
C.8	25
C.9	26
C.10	27
D.1	28
D.2	29

JOB SHEETS

HEATING AND AIR CONDITIONING JOB SHEET 1

Filling Out a Work Order

Name ______________________ Station ______________ Date ______________

NATEF Correlation:

This Job Sheet addresses the following NATEF task:

A.1. Complete work order to include customer information, vehicle identifying information, customer concern, related service history, cause, and correction.

Objective

Upon completion of this job sheet, you will be able to prepare a service work order based on customer input, vehicle information, and service history.

Tools and Materials

An assigned vehicle or the vehicle of your choice

Service work order or computer-based shop management package

Parts and labor guide

Work Order Source

Describe the system used to complete the work order. If a paper repair order is being used, describe the source.

PROCEDURE

1. Prepare the shop management software for entering a new work order or obtain a blank paper work order. Task Completed ☐
2. Enter customer information, including name, address, and phone numbers onto the work order. Task Completed ☐
3. Locate and record the vehicle's VIN. Task Completed ☐
4. Enter the necessary vehicle information, including year, make, model, engine type and size, transmission type, license number, and odometer reading. Task Completed ☐
5. Does the VIN verify that the information about the vehicle is correct? ________
6. Normally, you would interview the customer to identify his or her concerns. However to complete this job sheet, assume the only concern is that the A/C unit is not putting out cool enough air and the problem is low refrigerant levels. Make sure you include in the estimate diagnostic time, which does not identify any system leaks. Therefore, all you need to do is recharge the system.. Task Completed ☐

7. The history of service to the vehicle can often help diagnose problems as well as indicate possible premature part failure. Gathering this information from the customer can provide some of this information. For this job sheet assume the vehicle has not had a similar problem and was not recently involved in a collision. Service history is further obtained by searching files based on customer name, VIN, and license number. Check the files for any related service work. Task Completed ☐

8. Search for technical service bulletins on this vehicle that may relate to the customer's concern. Task Completed ☐

9. Based on the customer's concern, service history, TSBs, and your knowledge, what is the likely cause of this concern?

____

____

10. Enter this information onto the work order. Task Completed ☐

11. Prepare to make a repair cost estimate for the customer. Identify all parts that may need to be replaced to correct the concern. List these here.

____

____

12. Describe the task(s) that will be necessary to replace the part.

____

____

13. Using the parts and labor guide, locate the cost of the parts that will be replaced and enter the cost of each item onto the work order at the appropriate place for creating an estimate. Task Completed ☐

14. Now, locate the flat rate time for work required to correct the concern. List each task and with its flat rate time.

____

____

15. Multiply the time for each task by the shop's hourly rate and enter the cost of each item onto the work order at the appropriate place for creating an estimate. Task Completed ☐

16. Many shops have a standard amount they charge each customer for shop supplies and waste disposal. For this job sheet, use an amount of ten dollars for shop supplies. Task Completed ☐

17. Add the total costs and insert the sum as the subtotal of the estimate. Task Completed ☐

18. Taxes must be included in the estimate. What is the sales tax rate and does it apply to both parts and labor, or just one of these?

____

____

19. Enter the appropriate amount of taxes to the estimate, than add this to the subtotal. The end result is the estimate to give the customer. Task Completed ☐

20. By law, how accurate must your estimate be?

____

____

21. Generally speaking, the work order is complete and is ready for the customer's signature. However, some businesses require additional information; make sure you enter that information to the work order. On the work order there is a legal statement that defines what the customer is agreeing to. Briefly describe the contents of that statement.

____

____

Problems Encountered

____

____

____

Instructor's Comments

____

____

____

HEATING AND AIR CONDITIONING JOB SHEET 2

Identifying Problems and Concerns

Name ______________________ Station ______________ Date ______________

NATEF Correlation

This Job Sheet addresses the following NATEF task:

A.2. Identify and interpret heating and air conditioning concerns; determine necessary action.

Objective

Upon completion of this job sheet, you will be able to define heating and air conditioning system problems or concerns, prior to diagnosing or testing the systems.

Protective Clothing

Goggles or safety glasses with side shields

Describe the vehicle being worked on:

Year ________ Make ____________________ Model ________________

VIN ____________________ Engine type and size ______________________

PROCEDURE

1. Start the engine and describe how the engine seems to be running:

 ____

 ____

2. Move the temperature control on the dash to its full heat position. Describe the amount of heat coming from the heater in each of the fan positions.

 ____

 ____

3. Were there any unusual noises or smells when the heater fan was running? If so, describe them and when they occurred.

 ____

 ____

4. Turn the temperature control to the defrost mode. Describe the amount of heat coming from the defroster in each of the fan positions.

 ____

 ____

5. Were there any unusual noises or smells when the heater fan was running? If so, describe them and when they occurred.

____

____

6. Turn the temperature control to cool or the air conditioning position. Describe what the engine did when the A/C was turned on.

____

____

7. Describe how the air feels in each of the various fan positions.

____

____

8. Were there any unusual noises or smells when the A/C fan was running? If so, describe them and when they occurred.

____

____

9. Summarize your findings in detail; include both improper and proper system operation.

____

____

____

____

____

____

____

10. Based on the above, what are your suspicions and conclusions?

____

____

____

____

Problems Encountered

____

____

____

Instructor's Comments

____

____

____

HEATING AND AIR CONDITIONING JOB SHEET 3

Gathering Vehicle Information

Name ______________________________ Station __________________ Date __________________

NATEF Correlation

This Job Sheet addresses the following NATEF tasks:

A.3. Research applicable vehicle and service information, such as heating and air conditioning operation, vehicle service history, service precautions, and technical service bulletins.

A.4. Locate and interpret vehicle and major component identification numbers (VIN, vehicle certification labels, calibration labels).

Objective

Upon completion of this job sheet, you will be able to gather service information about a vehicle and its heating and air conditioning system.

Tools and Materials

Appropriate service manuals

Computer

Protective Clothing

Goggles or safety glasses with side shields

Describe the vehicle being worked on:

Year __________ Make ________________________ Model ______________________

VIN _______________________ Engine type and size ___________________________

PROCEDURE

1. Using the service manual or other information source, describe what each letter and number in the VIN for this vehicle represents.

   ____

   ____

2. Locate the Vehicle Emissions Control Information (VECI) label and describe where you found it.

   ____

   ____

3. Summarize what information you found on the VECI label.

   ____

   ____

4. While looking in the engine compartment did you find a label regarding the vehicle's air conditioning system? Describe where you found it.

____

____

5. Summarize the information contained on this label.

____

____

6. Using a service manual or electronic database, locate the information about the vehicle's air conditioning system. List the major components of the system and describe how the system's pressure is controlled.

____

____

____

____

____

7. Using a service manual or electronic database, locate and record all service precautions regarding the air conditioning system noted by the manufacturer.

____

____

____

____

8. Using the information that is available, locate and record the vehicle's service history.

____

____

____

____

9. Using the information sources that are available, summarize all Technical Service Bulletins for this vehicle that relate to the heating and air conditioning system.

____

____

____

____

Problems Encountered

Instructor's Comments

HEATING AND AIR CONDITIONING JOB SHEET 4

Conducting a Performance Test on an Air Conditioning System

Name ______________________ Station ______________ Date ______________

NATEF Correlation

This Job Sheet addresses the following NATEF task:

A.5. Performance test A/C system; diagnose A/C system malfunctions using principles of refrigeration.

Objective

Upon completion of this job sheet, you will be able to conduct an A/C system performance test and use theory to help diagnose basic problems within the system.

Tools and Materials

Service manual

Thermometer

Protective Clothing

Goggles or safety glasses with side shields

Describe the vehicle being worked on:

Year ________ Make __________________ Model __________________

VIN __________________ Engine type and size ______________________

PROCEDURE

1. Describe the type of air conditioning system that is on this vehicle:

   ____

   ____

   ____

   ____

2. Carefully inspect the condenser for damage and check for anything that may block air flow. Describe your findings:

   ____

   ____

3. Inspect the drive belts and hoses for the A/C system and describe your findings:

____

____

4. Check the mounting of the compressor and describe your findings.

____

____

5. Start the engine and operate the heating and air conditioning control in each of its positions. List any switch position that the system did not respond to correctly:

____

____

6. Keeping your hands away from the cooling fan, measure the temperature of the air in front of the condenser. What was your measurement?

____

7. Place the thermometer in the center duct in the dash and set the A/C control to its maximum position and the fan to its highest speed. Task Completed ☐

8. Raise the speed of the engine to about 1500 rpm and note the reading on the thermometer. What was it?

____

9. What reading would be satisfactory? Why? Explain why outside ambient temperature will affect the temperature of the air coming out of the duct.

____

____

____

____

10. Listen to the system and record any unusual sounds. Be sure to listen to the system while the compressor clutch cycles on and off.

____

____

11. Carefully place your hand on the condenser near the refrigerant inlet. Describe what you feel:

____

____

12. Carefully place your hand on the condenser near the refrigerant outlet. Describe what you feel:

____

____

13. Was there a change in temperature between the inlet and the outlet?

____

14. Based on the above inspection and performance test, what are your conclusions and recommendations?

____

____

____

____

Problems Encountered

____

____

____

Instructor's Comments

____

____

____

HEATING AND AIR CONDITIONING JOB SHEET 5

Conducting a Visual Inspection of an Air Conditioning System

Name ______________________ Station ______________ Date ______________

NATEF Correlation

This Job Sheet addresses the following NATEF tasks:

A.6. Diagnose abnormal operating noises in the A/C system; determine necessary action.

B.2.3. Inspect A/C condenser for airflow restrictions; perform necessary action.

Objective

Upon completion of this job sheet, you will be able to conduct a thorough visual inspection of an air conditioning system before beginning detailed testing of the system.

Tools and Equipment

Service manual

Feeler gauge set

Protective Clothing

Goggles or safety glasses with side shields

Describe the vehicle being worked on:

Year ________ Make ____________________ Model ____________________

VIN ____________________ Engine type and size ____________________

Describe the type of air conditioning used on the vehicle:

____

PROCEDURE

1. Check the condition of the compressor's drive belt. Describe its condition:

   ____

   ____

   ____

2. Check the refrigerant hose and fittings from the

   a. Compressor to the condenser

   b. Receiver/drier to accumulator

   c. Condenser to the evaporator

   d. Evaporator to compressor

Describe your findings:

___

___

___

3. Check all electrical connections to the system. Describe your findings:

___

___

___

4. Check the condenser by looking for any dirt or debris buildup that could cause decreased airflow through the condenser. Describe your findings:

___

___

___

5. Locate the compressor clutch air gap specifications. The specified gap is:

___

6. Measure the gap with a feeler gauge. The measured gap is: ___

If the gap is outside of specifications, what should you do?

___

___

___

7. Start the engine and turn on the air-conditioning system. Task Completed ☐

8. After the system has been on for a few minutes, check the condenser by feeling up and down the face or along the return bends for a temperature change. Describe your findings:

___

___

___

There should be a gradual change from hot to warm as you go from the top to the bottom. Any abrupt change indicates a restriction, and the condenser has to be flushed or replaced.

9. If the system has a receiver/drier, check it. The inlet and outlet lines should be the same temperature. Any temperature difference or frost on the lines or receiver tank is a sign of a restriction. Describe your findings:

___

___

___

10. If the system has a sight glass, check it. Describe your findings:

____

____

____

11. Feel the liquid line from the receiver/drier to the expansion valve. The line should be warm for its entire length. Describe your findings:

____

____

____

12. The expansion valve should be free of frost, and there should be a sharp temperature difference between its inlet and outlet. Describe your findings:

____

____

____

13. The suction line to the compressor should be cool to the touch from the evaporator to the compressor. If it is covered with thick frost, that might indicate that the expansion valve is flooding the evaporator. Describe your findings:

____

____

____

14. Check the entire system for the presence of frost. Typically the formation of frost on the outside of a line or component means there is a restriction to the flow of refrigerant. Describe your findings:

____

____

____

15. On vehicles equipped with the orifice tube system, feel the liquid line from the condenser outlet to the evaporator inlet. A restriction is indicated by any temperature change in the liquid line before the crimp dimples the orifice tube in the evaporator inlet. Describe your findings:

____

____

____

16. Based on the visual inspection, what are your conclusions about this air-conditioning system?

____

____

____

Problems Encountered

____

____

____

Instructor's Comments

____

____

____

HEATING AND AIR CONDITIONING JOB SHEET 6

Identifying the Type of Air-Conditioning System in a Vehicle

Name ______________________ Station ______________ Date ______________

NATEF Correlation

This Job Sheet addresses the following NATEF task:

A.7. Identify refrigerant type; select and connect proper gauge set; record pressure readings.

Objective

Upon completion of this job sheet, you will be able to use a service manual and the labeling in the engine compartment to identify the type of air conditioning used in a vehicle and check the system's overall performance.

Tools and Equipment

Service manual

Protective Clothing

Goggles or safety glasses with side shields

Describe the vehicle being worked on:

Year ________ Make ____________________ Model ____________________

VIN ____________________ Engine type and size ____________________

PROCEDURE

1. Look at the instrument panel controls for the air-conditioning system. Task Completed ☐
2. Describe the controls:

   ____

   ____

3. Look at all of the labeling in the engine compartment and locate any label that addresses the air conditioning system. Task Completed ☐
4. Summarize the contents of the labeling:

   ____

   ____

5. Look carefully at the fittings used at the compressor. What types of fittings are they?

   ____

   ____

   ____

6. What refrigerant is this system designed to use? ___________ What type of refrigerant oil must be used? ___________
7. Locate the information about this vehicle's air conditioning system in the service manual. Task Completed ☐
8. Record the service manual and the pages where the information was found.

____

____

____

9. Read through the material given in the service manual and determine what controls the flow of refrigerant through the evaporator. Summarize your findings here:

____

____

____

10. Give a complete description of this air conditioning system:

____

____

____

____

____

____

Problems Encountered

____

____

____

Instructor's Comments

____

____

____

HEATING AND AIR CONDITIONING JOB SHEET 7

Using a Pressure Gauge Set

Name ______________________ Station ______________ Date ______________

NATEF Correlation:

This Job Sheet addresses the following NATEF task:

A.7. Identify refrigerant type; select and connect proper gauge set; record pressure readings.

Objective

Upon completion of this job sheet, you will be able to correctly connect a pressure gauge set to an air conditioning system and record pressures.

Tools and Materials

Service manual

Manifold gauge set

Protective Clothing

Goggles or safety glasses with side shields

Describe the vehicle being worked on:

Year ________ Make ____________________ Model ____________________

VIN ____________________ Engine type and size ______________________

PROCEDURE

1. What type of refrigerant is required for this vehicle?

____

____

2. Gather the appropriate manifold and gauge set. What color are the hoses on the gauge set and what do the colors indicate?

____

____

3. Locate the low-side service valve and describe where it is.

____

____

4. Locate the high-side service valve and describe its location.

____

____

5. Make sure the valves on the gauge set are fully closed. Task Completed ☐
6. Remove the protective cap from the low-side service valve. Task Completed ☐
7. Connect the low-side service hose to the low-side service valve. Task Completed ☐
8. Remove the protective cap from the high-side service valve. Task Completed ☐
9. Connect the low-side service hose to the high-side service valve. Task Completed ☐
10. Start the engine and set to run at about 1500 rpm. How did you set this speed?

    ____

    ____

11. Place a fan in front of the radiator to provide adequate airflow. Task Completed ☐
12. Turn on the air conditioning system with all controls set to the maximum cooling position. Task Completed ☐
13. Observe the pressures on both gauges. What were they?

    ____

    ____

14. Compare your readings to specifications and describe what they indicate.

    ____

    ____

15. Return the engine to a normal idle speed. Task Completed ☐
16. Close the service hose valves, if so equipped. Task Completed ☐
17. Disconnect the service hoses. Task Completed ☐
18. Reinstall the protective caps on the service valves. Task Completed ☐
19. Turn off the A/C and the engine. Task Completed ☐

Problems Encountered

____

____

____

Instructor's Comments

____

____

____

HEATING AND AIR CONDITIONING JOB SHEET 8

Leak Testing an A/C System

Name ______________________ Station ______________ Date ______________

NATEF Correlation

This Job Sheet addresses the following NATEF task:

A.8. Leak test A/C system; determine necessary action.

Objective

Upon completion of this job sheet, you will be able to leak test an air conditioning system.

Tools and Materials

Leak detection dye

Ultraviolet black lamp

Electronic leak detector

Protective Clothing

Goggles or safety glasses with side shields

Describe the vehicle being worked on:

Year ________ Make ____________________ Model ____________________

VIN ____________________ Engine type and size ____________________

PROCEDURE

1. Inspect the system and look for oily residue on hoses, connections, and sealing points. Summarize the results of your check.

   ____

   ____

2. Inspect the system components for evidence of oil. Summarize the results of your check.

   ____

   ____

3. All suspected leak locations should be verified by using leak detection equipment. The following procedures are based on the two most common ways of leak detection. Follow the one that applies to the equipment you will be using. What equipment will you use to locate leaks?

   ____

   ____

Dye Leak Detection

1. Attach the manifold gauge set to the system's fittings. Task Completed ☐
2. Connect the center hose of the gauge set to one end of the dye injector and the other end to a can of refrigerant. Open the injector and add the refrigerant to the system. If the dye is previously mixed with refrigerant, add the mixture as a liquid to the system. Task Completed ☐
3. Operate the A/C system for at least 15 minutes, then turn off the A/C and the engine. Task Completed ☐
4. Using the ultraviolet black lamp, inspect all fittings and joints. Summarize the results of your check.

   ___

   ___

5. If there was no evidence of leakage, wait 24 hours and check again. Task Completed ☐

Electronic Leak Detection

1. Refer to the manufacturer's instructions and calibrate the leak detector. Task Completed ☐
2. Slowly move the probe of the detector along the refrigerant lines and A/C components. Keep the probe about one-quarter of an inch away from the surface of the lines and components. The beeping of the tester will increase as it gets closer to the leak. Summarize the results of your check.

   ___

   ___

3. If you detect a leak, verify the leak by using shop air to clear the immediate area around the suspected leak. Then test for leaks again. Summarize the results of your check. Task Completed ☐

Problems Encountered

___

___

___

Instructor's Comments

___

___

___

HEATING AND AIR CONDITIONING JOB SHEET 9

Working with Refrigerant Oils

Name ______________________ Station ______________ Date ______________

NATEF Correlation

This Job Sheet addresses the following NATEF tasks:

A.9. Inspect the condition of discharged oil; determine necessary action.

A.10. Determine recommended oil for system application.

Objective

Upon completion of this job sheet, you will be able to inspect the condition of discharged oil and select the correct type of oil for a system. You will also be able to measure and add oil to an A/C system.

Tools and Materials

Graduated container

Service manual

Protective Clothing

Goggles or safety glasses with side shields

Describe the vehicle being worked on:

Year ________ Make ____________________ Model ____________________

VIN ____________________ Engine type and size ____________________

PROCEDURE

1. The compressor oil level is checked only where there is evidence of a major loss of system oil that could be caused by a broken refrigerant hose, severe hose fitting leak, badly leaking compressor seal, or collision damage to the system's components. Is this the situation?

   ____

   ____

2. When replacing refrigerant oil, it is important to use the specific type and quantity of oil recommended by the compressor manufacturer. If there is a surplus of oil in the system, too much oil circulates with the refrigerant, causing the cooling capacity of the system to be reduced. Too little oil results in poor lubrication of the compressor. When there has been excessive leakage or it is necessary to replace a component of the refrigeration system, certain procedures must be followed to ensure that the total oil charge in the system is correct after leak repair or the new part is on the car. Is this the situation?

   ____

   ____

3. To maintain the original total oil charge, it is necessary to compensate for this by adding oil to the new replacement part. Because of the differences in compressor designs, be sure to follow the manufacturer's instructions when adding refrigerant oil to their unit. Refer to the service manual and give a summary of the amount of oil that should be added for each component that may be replaced.

 ____

 ____

 ____

 ____

4. Name the type of oil that should be used in this system. Normally, R-12 based systems use mineral oil, while R-134a systems use synthetic polyalkaline glycol (PAG) oils. Using a mineral oil with R-134a will result in A/C compressor failure because of poor lubrication. Use only the oil specified for the system.

 ____

5. If a component is being replaced and you are unsure of how much should be added to the system with the new component, completely drain the oil from the old component and measure it in a graduated container. Add that amount to the component when installing it. Task Completed ☐

Problems Encountered

____

____

____

Instructor's Comments

____

____

____

HEATING AND AIR CONDITIONING JOB SHEET 10

Using a Scan Tool on HVAC

Name ______________________ Station ______________ Date ______________

NATEF Correlation

This Job Sheet addresses the following NATEF task:

A.11. Using scan tool, observe and record related HVAC data and trouble codes.

Objective

Upon completion of this job sheet, you will be able to connect a scan tool and use the HVAC's control panel to retrieve diagnostic troubles codes and other HVAC related data.

Tools and Materials

Service manual

Scan tool

Protective Clothing

Goggles or safety glasses with side shields

Describe the vehicle being worked on:

Year ________ Make ____________________ Model ____________________

VIN ____________________ Engine type and size ____________________

PROCEDURE

Using a Scan Tool

1. Using the service manual or a component locator for this vehicle, locate where the diagnostic connector is for the body control module (BCM). Describe its location.

____

____

2. Connect the scan tool. Task Completed ☐
3. Turn the ignition on. Task Completed ☐
4. Program the scan tool for that vehicle. Task Completed ☐
5. From the menu on the scan tool, select the BCM. Task Completed ☐
6. Then retrieve and record all DTCs. What DTCs were observed and what do they indicate?

____

____

7. To observe the activity of the inputs and outputs, select DATA. Were any of the sensors or actuators operating out of their normal range? Which ones?

____

____

8. Turn the ignition off. Task Completed ☐

9. Disconnect the scan tool. Task Completed ☐

Using the Control Panel

1. Refer to the service manual for the correct procedure to enter into the HVACs self-diagnostic mode. Normally this is done by depressing and holding in buttons on the panel. What buttons are used on this vehicle?

____

____

2. Also, determine what display on the panel should be observed to retrieve the DTCs and data. Where will the display appear on the panel?

____

____

3. Turn the ignition on. Task Completed ☐

4. Press and hold the designated buttons for the designated time. How long should the buttons be held in for self-diagnostics?

____

____

5. Record all DTCs that were displayed and describe what is indicated by each.

____

____

6. If there are no DTCs detected and the system is still not operating properly, check the sensor input to the HVAC control unit. Task Completed ☐

7. The procedure for checking sensor inputs will also vary with the model of vehicle. Describe the procedure for doing this.

____

____

8. Turn the ignition switch OFF. Task Completed ☐

9. Depress and hold the designated buttons. Task Completed ☐

10. Start the engine and release the buttons after the engine starts. Task Completed ☐

11. The control panel should begin to display the data for the sensors. How is this information shown on this vehicle? Task Completed ☐

____

____

12. What is the procedure for canceling the sensor input display?

____

____

13. Turn the ignition switch OFF to cancel the self-diagnostic function. Task Completed ☐

14. After completing all repairs to the system, run the self-diagnostic test again to make sure there are no other problems. Task Completed ☐

Problems Encountered

____

____

____

Instructor's Comments

____

____

____

HEATING AND AIR CONDITIONING JOB SHEET 11

Inspect, Test, and Service a Compressor Clutch

Name ______________________ Station ______________ Date ______________

NATEF Correlation

This Job Sheet addresses the following NATEF tasks:

B.1.1. Diagnose A/C system conditions that cause the protection devices (pressure, thermal, and PCM) to interrupt system operation; determine necessary action.

B.1.3. Inspect, test, and/or replace A/C compressor clutch components and/or assembly.

Objective

Upon completion of this job sheet, you will be able to inspect, test, and service a compressor clutch.

Tools and Materials

Fender covers

Hand tools

Necessary pullers and installers

Test light

DMM

Service manual

Protective Clothing

Goggles or safety glasses with side shields

Describe the vehicle being worked on:

Year ________ Make ____________________ Model ____________________

VIN ____________________ Engine type and size ____________________

Describe the type of air conditioning found on the vehicle:

____

PROCEDURE

1. Place fender covers over the front fenders for protection. Task Completed ☐
2. Visually inspect the clutch for damage. Task Completed ☐
3. Check the electrical connections to the clutch. Task Completed ☐
4. Using a test light or DMM with the AC system on, check the electrical plug at the clutch for power. Task Completed ☐
5. If there is no power, locate and repair the problem. Retest. Task Completed ☐

6. When power is present, check the ground circuit with a DMM. Task Completed ☐
7. If it is determined that both power and ground are present at the compressor clutch, the clutch is defective and must be replaced. Task Completed ☐
8. On some vehicles, the clutch can be serviced on the vehicle. On others, the compressor assembly must be removed to service the clutch. Refer to the service manual on your specific vehicle to determine what must be done. Task Completed ☐
9. Using the proper pullers for your specific application and following the procedure in the service manual, remove the clutch. Task Completed ☐
10. Reverse the procedure and install the new clutch. Task Completed ☐
11. Test clutch operation after repairs are completed. Task Completed ☐

Problems Encountered

___

___

___

Instructor's Comments

___

___

___

HEATING AND AIR CONDITIONING JOB SHEET 12

Servicing Drive Belts

Name ______________________ Station ______________ Date ______________

NATEF Correlation

This Job Sheet addresses the following NATEF task:

B.1.2. Inspect and replace A/C compressor drive belts; determine necessary action.

Objective

Upon completion of this job sheet, you will be able to inspect, replace, and adjust the drive belt for an A/C compressor.

Tools and Materials

Service manual
Belt tension gauge
Straightedge
Pry bar

Protective Clothing

Goggles or safety glasses with side shields

Describe the vehicle being worked on:

Year ________ Make ____________________ Model ____________________

VIN ____________________ Engine type and size ____________________

Describe:

____

____

____

PROCEDURE

Serpentine Belts

1. Check the condition of the drive belt. Carefully look to see if it is cracked or glazed. What is the condition of the belt?

   ____

   ____

2. If the belt needs to be replaced, disconnect the electric cooling fan at the radiator, if the vehicle has one. Task Completed ☐

3. Always use the exact replacement size of belt. The size of a new belt is typically given, along with the part number, on the belt container. What is the part number?

   ____

4. Locate a belt routing diagram in a service manual or on an underhood decal. Where did you find this diagram?

   ____

5. Compare the diagram with the routing of the old belt. If the actual routing is different from the diagram, draw the existing routing here.

6. Loosen the tension on the old belt and remove the belt. How did you relieve the tension?

   ____

7. Inspect the grooves of the drive pulleys for rust, oil, wear, and other damage. What did you find and what service do you recommend?

   ____

   ____

8. Check the alignment of the pulleys. What service do you recommend?

   ____

   ____

9. Make sure the belt tensioner or idler pulley is working properly. This pulley may be a spring-loaded tensioner or an adjustable pulley. Describe its condition.

   ____

   ____

10. After all the parts are determined to be in good order, install the new belt. Make sure to wrap it according to instructions. Also, make sure the ribs of the belt are seated in the matched grooves on the pulleys. Task Completed ☐

11. Once the belt is fully routed, put tension on the belt and adjust it to specifications. What are the specifications?

    ____

    ____

V-Belts

1. Visually inspect the drive belt and pulleys for signs of cracks, wear, or breaks. What did you find?

____

____

2. Using a straightedge, check the alignment of the pulleys and describe what you found.

____

____

3. Loosen all of the mounting brackets for compressor. Task Completed ☐
4. Loosen all brackets on accessories that may have a belt in front of the compressor belt. Task Completed ☐
5. Remove the belt from the compressor. Task Completed ☐
6. Compare the new belt with the old one. Task Completed ☐
7. Install the new belt(s). Task Completed ☐
8. Use a pry bar to move the accessories and to tighten the belt. Make sure the pry bar is not against anything that will bend or break. Task Completed ☐
9. Check the service manual for belt tension specifications. What is the specification?

____

____

10. Adjust the belt to specifications. Task Completed ☐
11. Tighten all brackets that were loosened or removed. Task Completed ☐

Problems Encountered

____

____

____

Instructor's Comments

____

____

____

HEATING AND AIR CONDITIONING JOB SHEET 13

R&R a Compressor and Its Mountings

Name ______________________ Station ______________ Date ______________

NATEF Correlation

This Job Sheet addresses the following NATEF task:

B.1.4. Remove, inspect, and reinstall A/C compressor and mountings; determine required oil quantity.

Objective

Upon completion of this job sheet, you will be able to remove and replace an A/C compressor and mountings.

Tools and Materials

Refrigerant line caps

Line wrenches

Graduated container

Protective Clothing

Goggles or safety glasses with side shields

Describe the vehicle being worked on:

Year ________ Make ____________________ Model ____________________

VIN ____________________ Engine type and size ____________________

PROCEDURE

1. Disconnect the negative battery cable. Task Completed ☐
2. Identify and disconnect all electrical connections to the compressor. Task Completed ☐
3. Discharge the system using a recovery/recycling machine. Task Completed ☐
4. Disconnect the refrigerant lines at the compressor. Task Completed ☐
5. Immediately cap or seal the ends of the lines or hoses. Task Completed ☐
6. Remove the drive belt for the compressor. Task Completed ☐
7. Loosen and remove the compressor mounting brackets. Note the location of the bolts, as they are typically a different length. Task Completed ☐
8. Remove the compressor. Task Completed ☐
9. Clean the outside of the compressor. Task Completed ☐

10. Drain the oil from the compressor into the graduated container. Allow at least ten minutes for the oil to drain. How much oil came out?

____

____

11. Using the service manual, identify the type of oil and the amount that should be installed in the compressor.

____

____

12. Does the amount drained from the compressor equal the amount specified? If it is less, what is indicated?

____

____

13. Discard the old refrigerant oil according to the local and state requirements. What are these?

____

____

14. Add new oil in the correct amount as removed to the new or repaired compressor. Often, rebuild and new compressors are pre-filled with the correct oil, therefore no oil should be added. Did you need to add oil to the compressor?

____

____

15. To install a compressor, reverse the removal procedure. Make sure the drive belt is tightened to the proper tension. Task Completed ☐

16. Recharge the system with refrigerant. Task Completed ☐

17. Does the air conditioning system now perform satisfactorily?

____

____

Problems Encountered

____

____

____

Instructor's Comments

____

____

____

HEATING AND AIR CONDITIONING JOB SHEET 14

Hybrid Vehicle A/C Systems

Name ______________________ Station ______________ Date ______________

NATEF Correlation:

This Job Sheet addresses the following NATEF task:

B.1.5. Identify hybrid vehicle AC system electrical circuits, service and safety precautions.

Objective

Upon completion of this job sheet, you will be able to locate and safely disconnect the high voltage circuit on a hybrid vehicle and safely work on and around the air conditioning system.

Tools and Materials

Service manual
Hybrid vehicle
Vinyl tape for insulation

Protective Clothing

Goggles or safety glasses with side shields
Insulated gloves that are dry and are not cracked, ruptured, torn, or damaged in any way.

Describe the vehicle being worked on:

Year ________ Make ____________________ Model ____________________

VIN ____________________ Engine type and size ____________________

PROCEDURE

WARNING: *Unprotected contact with any electrically charged ("hot" or "live") high-voltage component could cause serious injury or death.*

1. Describe how this vehicle is identified as a hybrid.

____

____

____

2. After referring to the vehicle's owner's manual, briefly describe its operation.

____

____

____

____

3. How many volts are the batteries rated at?

____

4. Where are the batteries located?

____

____

5. How are the high-voltage cables labeled and identified?

____

____

6. What is used to provide short circuit protection in the high-voltage battery pack?

____

____

7. What isolates the high-voltage system from the rest of the vehicle when the vehicle is shut off?

____

____

8. Without touching the high-voltage cables or components, describe the routing of the cables. Include what they appear to be connected from and to.

____

____

____

____

WARNING: *Never assume that a hybrid vehicle is shut off simply because it is silent. Make sure the ignition key is in your pocket and not in the ignition switch.*

9. Describe the procedure for totally isolating the high-voltage system from the vehicle. This often involves the removal of a service plug. Be sure to include the location of this plug in your description of the procedure.

____

____

____

____

____

____

____

____

10. Make two signs saying, "WORKING ON HIGH VOLTAGE PARTS. DO NOT TOUCH!" Attach one to the steering wheel, and set the other one near the parts you are working on. Task Completed ☐
11. After the service plug has been removed, how long should you wait before working around or on the high-voltage system?

____

____

12. Describe the type of drive the A/C compressor uses:

____

____

13. Name the type of oil that must be installed in this compressor and explain why using the correct type is critical.

____

____

14. If the wrong type of oil has been installed in the compressor or A/C system, what must be done?

____

____

15. Refer to the service manual and list at least five precautions that must be followed when servicing this A/C system.

____

____

____

____

____

____

____

16. To inspect or replace the compressor, the high-voltage system of the vehicle must be isolated. Task Completed ☐

    CAUTION: Because the compressor has a high-voltage circuit, wear insulated gloves and isolate the high-voltage circuit before inspection or service.

17. To remove the compressor, disconnect the power cables to the compressor and wrap the terminal ends with insulating tape before removing it from its mount. Task Completed ☐
18. After service, reconnect the battery and the high voltage circuit. Task Completed ☐

19. Some systems may need to be reinitialized, what are they?

____

____

____

Problems Encountered

____

____

____

Instructor's Comments

____

____

____

HEATING AND AIR CONDITIONING JOB SHEET 15

In-Line Filter Installation

Name ______________________ Station ______________ Date ______________

NATEF Correlation

This Job Sheet addresses the following NATEF task:

B.2.1. Determine need for an additional A/C system filter; perform necessary action.

Objective

Upon completion of this job sheet, you will be able to determine whether or not the A/C system needs a new filter and be able to correctly install one if necessary.

Tools and Materials

In-line filter assembly

Line wrenches

Service manual

Protective Clothing

Goggles or safety glasses with side shields

Describe the vehicle being worked on:

Year ________ Make ____________________ Model ____________________

VIN ____________________ Engine type and size ____________________

PROCEDURE

To determine if the system needs an in-line filter, answer the following questions:

1. Did the compressor need to be replaced because it was seized? __________
2. Does the system have a history of plugged metering devices? __________
3. If the answer to either of these questions was yes, the system needs to have a filter installed. Why would these conditions warrant a supplemental filter?

____

____

____

4. Describe the in-line filter: does it have an orifice or is it just a filter?

____

____

____

5. If the filter does not have an orifice, locate a convenient location for installation in the liquid line between the condenser outlet and the evaporator inlet. Describe where you will install the filter.

____

____

____

6. If the filter has an orifice, find a convenient location for installation in the low-pressure side of the system beyond the expansion tube. Describe where you will install the filter.

____

____

____

7. Make sure the system is void of all refrigerant. How did you do this?

____

____

____

8. Secure the necessary tools, and then prepare the filter for installation according to the instructions that accompany the filter. What did you need to do?

____

____

____

9. Install the filter into the appropriate line. Task Completed ☐

10. If the filter had a built-in orifice, remove the original expansion tube from the system. Task Completed ☐

11. Make sure the lines are properly sealed, and then prepare to charge the system. What did you need to do before charging the system?

____

____

____

____

Problems Encountered

Instructor's Comments

HEATING AND AIR CONDITIONING JOB SHEET 16

R&R A/C System Components and Hoses

Name ______________________ Station ______________ Date ______________

NATEF Correlation

This Job Sheet addresses the following NATEF tasks:

B.2.2. Remove and inspect A/C system mufflers, hoses, lines, fittings, O-rings, seals, and service valves; perform necessary action.

B.2.4. Remove, inspect, and install receiver/drier or accumulator/drier; determine required oil quantity.

B.2.5. Remove and install expansion valve or orifice (expansion) tube.

Objective

Upon completion of this job sheet, you will be able to remove and install various A/C components, including hoses and lines.

Tools and Materials

Manifold gauge set
Refrigerant recovery equipment
Vacuum pump
Orifice tube remover/installer
Hacksaw
Flare nut wrenches
Razor blades
Spring lock tool

Protective Clothing

Goggles or safety glasses with side shields

Describe the vehicle being worked on:

Year __________ Make ____________________ Model ____________________

VIN ____________________ Engine type and size ____________________

PROCEDURE

1. Purge and recover the refrigerant in the system. Task Completed ☐
2. Disconnect the negative cable of the battery. Task Completed ☐
3. Based on your diagnosis of the system, what components need to be replaced?

____

____

____

Refrigerant Oil

1. Whenever you replace a component, drain the oil from the component, after it has been removed, into a graduated container. Record the amount of oil drained from each of the parts in this job sheet.

   a. Hoses

   ____

   b. Receiver/Drier

   ____

   c. Accumulator

   ____

   d. Thermostatic Expansion Valve

   ____

   e. Orifice Tube

   ____

2. What type of oil is specified for this A/C system?

   ____

3. According to the service manual, how much oil should be added to the system when each of the following components is replaced?

   a. Hoses

   ____

   b. Receiver/Drier

   ____

   c. Accumulator

   ____

   d. Thermostatic Expansion Valve

   ____

   e. Orifice Tube

   ____

Hoses and Fittings

1. Although total hose replacement is the preferred way to correct for a hose leak, there are several accepted ways to repair refrigerant hoses and fittings. Using insert barb fittings and a length of replacement hose, you can fabricate an acceptable replacement for an original equipment hose. Insert barb fittings can also be used to replace the damaged section of a hose. The process for doing this is the same as what follows. To do this, measure and mark the required length of replacement high-pressure hose.

   Task Completed ☐

2. Using the razor blade, cut the hose to the desired length. Task Completed ☐
3. Examine the cut end and trim as necessary to have a square and flat cut. Task Completed ☐
4. Apply clean refrigerant oil to the inside of the hose. Task Completed ☐
5. Coat the fitting's insert with refrigerant oil and insert it into the hose. Task Completed ☐
6. Install the hose clamp and position it so that it is over the barb closest to the fitting. Then tighten it. Task Completed ☐
7. When using an insert barb fitting to replace a bad original equipment fitting, the original fitting must be removed without damaging too much of the hose. Remove the hose from the vehicle. Task Completed ☐
8. Set the hose in a vise to hold it while cutting through the fitting's ferrule with a hacksaw. Make this cut in the direction of the hose, not through the hose. Task Completed ☐
9. With the ferrule cut, use pliers to peel or pull the ferrule off. Task Completed ☐
10. Use the razor blade to cut the hose at a point just beyond the insert of the original fitting. Task Completed ☐
11. Examine the cut end and trim as necessary to have a square and flat cut. Task Completed ☐
12. Apply clean refrigerant oil to the inside of the hose. Task Completed ☐
13. Coat the fitting's insert with refrigerant oil and insert it into the hose. Task Completed ☐
14. Install the hose clamp and position it so that it is over the barb closest to the fitting. Then tighten it. Task Completed ☐
15. If the hose or line has spring lock fittings, a special tool is required to separate the sections. Put the tool over the coupling. Task Completed ☐
16. Close the tool and push the lines into the tool to release the female fitting from the garter spring of the coupling. Task Completed ☐
17. Pull the male and female fittings apart. Task Completed ☐
18. Remove the tool. Task Completed ☐
19. When joining two fittings with a spring lock coupling, lubricate new O-rings and put them in their proper location on the male fitting. Task Completed ☐
20. Insert the male fitting into the female fitting. Task Completed ☐
21. Firmly push them together until they are secured by the spring lock coupling. Task Completed ☐

Receiver/Drier

1. If the unit has a low-pressure switch, disconnect the wires to it. Task Completed ☐
2. Remove the inlet and outlet hoses at the receiver/drier. Task Completed ☐
3. Disconnect the mounting bolts and brackets from the receiver/drier. Task Completed ☐
4. Lift the receiver/drier out. Task Completed ☐
5. The procedure for installation is the reverse of removal. Task Completed ☐

Accumulator

1. Disconnect the lines at the accumulator's inlet and outlet fittings. Task Completed ☐
2. Loosen and/or remove the accumulator mounting bolts or screws. Task Completed ☐
3. Lift the accumulator out. Task Completed ☐
4. The procedure for installation is the reverse of removal. Task Completed ☐

Thermostatic Expansion Valve (TXV)

1. Disconnect the inlet and outlet lines at the TXV. Task Completed ☐
2. Remove whatever is used to keep the remote sensing bulb secure. Task Completed ☐
3. Loosen and/or remove the mounting clamp for the TXV. Task Completed ☐
4. Remove the TXV from the evaporator. Task Completed ☐
5. The procedure for installation is the reverse of removal. Task Completed ☐

Orifice Tube

1. Disconnect the inlet line at the evaporator. Task Completed ☐
2. Pour a small amount of clean refrigerant oil into the orifice tube. Task Completed ☐
3. Insert the orifice removal tool. Task Completed ☐
4. Turn the handle of the tool just enough for it to engage onto the tabs of the orifice tube. Task Completed ☐
5. Hold the handle of the tool in position while turning the tool's outer sleeve clockwise to remove the orifice tube. Task Completed ☐
6. Coat the new orifice tube with clean refrigerant oil. Task Completed ☐
7. Place the orifice tube into the evaporator line inlet. Push it in until it stops. Task Completed ☐
8. Install a new O-ring on the refrigerant line and reconnect it. Task Completed ☐

Problems Encountered

____

____

____

Instructor's Comments

____

____

____

HEATING AND AIR CONDITIONING JOB SHEET 17

Servicing the Evaporator Water Drain

Name ______________________ Station ______________ Date ______________

NATEF Correlation

This Job Sheet addresses the following NATEF task:

B.2.6. Inspect evaporator housing water drain; perform necessary action.

Objective

Upon completion of this job sheet, you will be able to inspect the evaporator housing water drain.

Tools and Materials

Drain pan
Clean rags
Air nozzle

Protective Clothing

Goggles or safety glasses with side shields

Describe the vehicle being worked on:

Year ________ Make ____________________ Model ____________________

VIN ____________________ Engine type and size ____________________

PROCEDURE

1. Raise the vehicle on a lift. Task Completed ☐
2. Locate the evaporator case drain tube and describe its overall condition.

   ____

   ____

3. Place the drain pan under the evaporator case. Task Completed ☐
4. Disconnect the tube from the case. Did water come out of the case? ________
5. If no water came out, carefully clean out the drain hole and tube at the evaporator case. If it is necessary to insert a rod through the tube to clean it out, do this carefully because the tube is accordion shaped and one of its bends may feel like a restriction. Task Completed ☐

6. If no water came out when you attempted to clear this tube, the entire evaporator case must be cleared. Describe your results so far.

____

____

7. If water came out of the case, the outside drain tube should be cleared by inserting the air nozzle into an end of the tube. Low air pressure should remove any restrictions in the tube. Describe what happened.

____

____

Problems Encountered

____

____

____

Instructor's Comments

____

____

____

HEATING AND AIR CONDITIONING JOB SHEET 18

Removing and Installing an Evaporator

Name ______________________ Station ______________ Date ______________

NATEF Correlation

This Job Sheet addresses the following NATEF task:

B.2.7. Remove, inspect, and reinstall evaporator; determine required oil quantity.

Objective

Upon completion of this job sheet, you will be able to correctly remove and reinstall an evaporator. You will also be able to determine the amount of oil that should be added to the system after a new evaporator has been installed.

Tools and Materials

Charging station
Gauge set
Graduated container
Hand tools
Miscellaneous plugs and hoses
New refrigerant oil
Refrigerant recovery/recycling machine
Service manual

Protective Clothing

Goggles or safety glasses with side shields

Describe the vehicle being worked on:

Year ________ Make ________________ Model ________________

VIN ________________ Engine type and size ________________

PROCEDURE

1. Recover the refrigerant using an approved recovery/recycling equipment. Task Completed ☐
2. Disconnect the negative battery cable. Task Completed ☐
3. If the evaporator and heater core are a combined unit, drain the engine's coolant. Task Completed ☐
4. Disconnect and label all electrical connectors, cables, and vacuum hoses that are connected to the evaporator. What did you need to disconnect?

   ____

   ____

5. Disconnect the refrigerant hoses at the evaporator and plug or cap the hose ends to prevent dirt and moisture from entering the system. Task Completed ☐
6. Unbolt and remove the evaporator. What did you need to remove to gain access to the evaporator?

   ____

   ____

7. Drain the oil from the evaporator into a graduated container. How much oil was drained from the evaporator?

____

8. Check the oil for dirt. If the oil is contaminated, replace the accumulator or receiver/drier. Describe the condition of the oil.

____

____

9. Add the above amount of new refrigerant oil or the amount specified in the service manual to the new evaporator. How much oil did you put into the evaporator?

____

____

10. Coat the new O-rings for the evaporator and the line fittings on the evaporator with clean refrigerant oil. Task Completed ☐

11. Install the new evaporator and tighten the fittings. Task Completed ☐

12. Reinstall and reconnect all parts, wires, cables, and vacuum hoses that were disconnected during removal. Add coolant if needed. Task Completed ☐

13. Connect the negative battery cable. Task Completed ☐

14. Evacuate and recharge the system. How much refrigerant was put in the system?

____

15. Perform a leak test and correct any problems. Were there any problems?

____

____

16. Drive the vehicle and then conduct a performance test. Does the system work as it should? Explain.

____

____

____

____

Problems Encountered

Instructor's Comments

HEATING AND AIR CONDITIONING JOB SHEET 19

Removing and Installing a Condenser

Name ______________________ Station ______________ Date ______________

NATEF Correlation

This Job Sheet addresses the following NATEF task:

B.2.8. Remove, inspect, and reinstall condenser; determine required oil quantity.

Objective

Upon completion of this job sheet, you will be able to correctly remove and reinstall a condenser. You will also be able to determine the amount of oil that should be added to the system after a new condenser has been installed.

Tools and Materials

Charging station
Gauge set
Graduated container
Hand tools
Miscellaneous plugs and hoses
New refrigerant oil
Refrigerant recovery/recycling machine
Service manual

Protective Clothing

Goggles or safety glasses with side shields

Describe the vehicle being worked on:

Year ________ Make ________________ Model ________________

VIN ________________ Engine type and size ________________

PROCEDURE

1. Recover the refrigerant using an approved recovery/recycling equipment. Task Completed ☐
2. Disconnect the negative battery cable. Task Completed ☐
3. On many vehicles, the radiator, cooling fan, and/or shroud must be removed to gain access to the condenser. If the radiator must be removed, drain the coolant. What do you need to remove?

   ____

   ____

4. Disconnect and label all electrical connectors that are connected to the condenser and all those that may be in the way. What did you need to disconnect?

   ____

   ____

5. Disconnect the refrigerant hoses at the condenser and plug or cap the hose ends to prevent dirt and moisture from entering the system. Task Completed ☐

6. Unbolt and remove the evaporator. If the receiver/drier mounts to the condenser, disconnect it and remove it with the condenser. Task Completed ☐

7. Drain the oil from the condenser into a graduated container. How much oil was drained from the condenser?

   ____

8. Check the oil for dirt. If the oil is contaminated, replace the accumulator or receiver/drier. Describe the condition of the oil.

   ____

   ____

9. Add the above amount of new refrigerant oil or the amount specified in the service manual to the new condenser. How much oil did you put into the evaporator?

   ____

   ____

10. Install the new condenser and loosely tighten the mounting bolts. Task Completed ☐

11. Coat the new O-rings for the condenser and the line fittings on the condenser with clean refrigerant oil. Task Completed ☐

12. Tighten the fittings, then tighten the mounting bolts. Task Completed ☐

13. Reinstall and reconnect all parts, wires, cables, and vacuum hoses that were disconnected or removed during removal. Add coolant if needed. Task Completed ☐

14. Connect the negative battery cable. Task Completed ☐

15. Evacuate and recharge the system. How much refrigerant was put in the system?

    ____

16. Perform a leak test and correct any problems. Were there any problems?

    ____

    ____

17. Drive the vehicle and then conduct a performance test. Does the system work as it should? Explain.

    ____

    ____

    ____

    ____

Problems Encountered

Instructor's Comments

HEATING AND AIR CONDITIONING JOB SHEET 20

Basic Heater Inspection and Checks

Name ______________________ Station ______________ Date ______________

NATEF Correlation

This Job Sheet addresses the following NATEF task:

C.1. Diagnose temperature control problems in the heater/ventilation system; determine necessary action.

Objective

Upon completion of this job sheet, you will be able to diagnose temperature control problems in the heater/ventilation system.

Tools and Materials

Pyrometer

Protective Clothing

Goggles or safety glasses with side shields

Describe the vehicle being worked on:

Year ________ Make ____________________ Model ____________________

VIN ____________________ Engine type and size ____________________

PROCEDURE

1. When there is a problem of insufficient heat, begin your diagnosis with a visual inspection and a check of the coolant level. Summarize your findings.

 ____

 ____

2. Check the engine's cooling system for proper operation and signs of leaks. Summarize your findings.

 ____

 ____

3. If the cooling system seems to be fine, check the operation of the blower motor. Summarize your findings.

 ____

 ____

4. Then turn the heater controls on, and run the engine until it reaches normal operating temperature. Then measure the temperature of the upper radiator hose. The temperature can be measured with a pyrometer. If one is not available, gently touch the hose. You should not be able to hold the hose long because of the heat. While doing this, make sure you stay clear of the area around the cooling fan. If the temperature of the hose is not within specifications, suspect a faulty thermostat. Summarize your findings.

____

____

5. If the hose was the correct temperature, check the temperature of the two heater hoses. They should both be hot. If only one of the hoses is hot, suspect the heater control valve or a plugged heater core. Summarize your findings.

____

____

6. Most often, if these two items are faulty, the engine's cooling system will be negatively affected. Both of these items are replaced, rather than repaired. Summarize your findings.

____

____

Problems Encountered

____

____

____

Instructor's Comments

____

____

____

HEATING AND AIR CONDITIONING JOB SHEET 21

Test Cooling System

Name ______________________ Station ______________ Date ______________

NATEF Correlation

This Job Sheet addresses the following NATEF tasks:

C.2. Perform cooling system pressure tests; check coolant condition, inspect and test radiator, pressure cap, coolant recovery tank, and hoses; perform necessary action.

C.5. Determine coolant condition and coolant type for vehicle application; drain and recover coolant.

C.6. Flush system; refill system with recommended coolant; bleed system.

Objective

Upon completion of this job sheet, you will be able to perform tests on the cooling system, cap, and recovery system pressure; and temperature, as well as test, drain, and recover coolant. You will also be able to flush and refill the cooling system with the recommended coolant and bleed the air as required.

Tools and Materials

Clean cloth rags
Cooling system tester
Coolant hydrometer
Drain pan
Hand tools
Cooling system flusher
Coolant recycler

Protective Clothing

Goggles or safety glasses with side shields

Describe the vehicle being worked on:

Year ________ Make ____________________ Model ____________________

VIN ____________________ Engine type and size ____________________

Describe general condition:

____

PROCEDURE

1. Wipe off the radiator filler neck and inspect it. Task Completed ☐

   Is the sealing seat free of accumulated dirt, nicks, or anything that might prevent a good seal? Yes ☐ No ☐

Are the cams on the out-turned flange bent or worn? Yes ☐ No ☐

Not Applicable ☐

2. Inspect the tube from the radiator to the expansion tank for dents and other obstructions. Run wire through the tube to be certain it is clear. Task Completed ☐

3. To test the cooling system for external leaks, attach a cooling system tester to the appropriate flexible adapter and carefully pump up pressure. Looking at tester's gauge, bring the pressure up to the proper cooling system test point indicated on the dial face. Task Completed ☐

   Does the pressure begin to drop? Yes ☐ No ☐

   If so, check all external connections, including hoses, gaskets, and heater core for leaks. Task Completed ☐ Not Applicable ☐

4. Check all points closely for small pinhole leaks or potentially dangerous weak points in the system. Task Completed ☐

   To test for internal leaks:

5. Remove the flexible adapter from radiator, and replace the pressure cap. Task Completed ☐

6. Start the engine and allow it to reach normal operating temperature so the thermostat will open fully. Task Completed ☐

7. Slowly and carefully remove the pressure cap and replace it with the flexible adapter. Task Completed ☐

   WARNING: *Engine coolant will be under pressure. Always wear hand protection and safety goggles or glasses with side shields when doing this task.*

8. Lock the tester slowly into place. Watch for pressure buildup. Task Completed ☐

   WARNING: *If the pressure builds up suddenly due to an internal leak, remove the tester immediately.*

9. If no immediate pressure buildup is visible on the gauge, keep the engine running. Pressurize the cooling system. Task Completed ☐ Not Applicable ☐

10. Does the gauge fluctuate? If so, the cooling system has a compression or combustion leak. Yes ☐ No ☐

11. Before removing it, hold the tester body and press the pressure release button against some object on the car to relieve the pressure in the system. Task Completed ☐ Not Applicable ☐

    WARNING: *Always wear hand protection and safety goggles or glasses with side shields when doing this task. Avoid the hot coolant and steam.*

12. After the pressure has been relieved, shield your hands with a cloth wrapped around the flexible adapter and filler neck. Slowly turn the adapter cap from the lock position to the safety unlock position of the radiator filler neck cam. Do not detach the flexible adapter from the filler neck. Allow the pressure to dissipate completely in this safety position. Then you can remove the adapter and reinstall the radiator cap. Task Completed ☐ Not Applicable ☐

13. Summarize your findings from these checks.

____

14. Draw some coolant out of the radiator or recovery tank with the coolant hydrometer. Task Completed ☐

15. Check the strength of the coolant by observing the position of the tester's float. Summarize what you found out about the coolant and what service you recommend.

____

____

16. Reinstall the radiator cap, then start the engine and move the heater control to its full heat position. Task Completed ☐

17. Turn off the engine after it has only slightly warmed up. Task Completed ☐

18. Place a drain pan under the radiator drain plug. Task Completed ☐

19. Make sure the engine and cooling system is not hot, and then open the drain plug. Keep the radiator cap on until the recovery tank is emptied. Task Completed ☐

20. Once all coolant has drained, close the drain plug. Then empty the drain pan with the coolant into the coolant recycler and process the used coolant according to the manufacturer's instructions. Task Completed ☐

21. Connect the flushing machine to the cooling system. Task Completed ☐

22. Follow the manufacturer's instructions for machine operation. Typically, flushing continues until clear water flows out of the system. Once flushing is complete, disconnect the flushing machine. Task Completed ☐

23. What type of coolant is recommended for this vehicle? Where did you find the recommendation?

____

____

24. Look up the capacity of the cooling system in the service manual.

    What is it? ______________________________

25. Prepare to add a mixture of 50% coolant and 50% water. Task Completed ☐

26. Locate the engine's cooling system bleed valve. Where is it?

____

27. Loosen the bleed valve. Task Completed ☐

28. Pour an amount of coolant that is equal to half of the cooling system's capacity into the system. Task Completed ☐

29. Add water until some coolant begins to leak from the bleed valve. Close the valve. Task Completed ☐

30. Leave the radiator cap off and start the engine. Task Completed ☐

31. Continue to add water to the system as the engine warms up. Task Completed ☐

32. Once the system appears full, install the radiator cap. Task Completed ☐

33. Observe the system for leaks and watch the activity in the recovery tank. If no coolant moves to the tank, the system may need to be bled. Task Completed ☐

Problems Encountered

____

____

____

Instructor's Comments

____

____

____

HEATING AND AIR CONDITIONING JOB SHEET 22

Checking Cooling System Hoses and Belts

Name ______________________ Station ______________ Date ______________

NATEF Correlation

This Job Sheet addresses the following NATEF task:

C.3. Inspect engine cooling and heater system hoses and belts; perform necessary action.

Objective

Upon completion of this job sheet, you will be able to inspect and replace cooling and heater system hoses and inspect the drive belt for the A/C comperssor.

Tools and Materials

Basic hand tools
Straightedge
Knife
Wire brush
Emery cloth
Belt tension gauge

Protective Clothing

Goggles or safety glasses with side shields

Describe the vehicle being worked on:

Year ________ Make ____________________ Model ____________________

VIN ____________________ Engine type and size ____________________

PROCEDURE

Belts

1. Check the condition of the drive belts and describe what you found here.

   ____

   ____

2. What is the specified tension for the drive belt(s)?

   ____

   ____

3. Check the tension of the belt(s) on the vehicle. What was it?

   ____

   ____

4. What services do you recommend?

____

____

Hose Service

1. Carefully check all cooling hoses for leakage, swelling, and chafing. Record your findings here.

____

____

____

2. Squeeze each hose firmly. Do you notice any cracks or signs of splits when hoses are squeezed? Do any hoses feel mushy or extremely brittle?

____

____

____

3. Carefully examine the areas around the clamps. Are there any rust stains? What is or would be indicated by rust stains?

____

____

____

4. Carefully check the lower radiator hose. This hose contains a coiled wire lining to keep it from collapsing during operation. If the wire loses tension, the hose can partially collapse at high speed and restrict coolant flow. Describe this hose's condition.

____

____

5. Identify all hoses that need to be replaced.

____

____

____

____

____

6. Gather the new hoses, hose clamps, and coolant. Task Completed ☐

7. Release the pressure in the system by slowly and carefully opening the radiator cap. Task Completed ☐

8. Once the pressure is relieved, drain the coolant system below the level that is being worked on. Make sure the drained coolant is collected and recycled or disposed of according to local regulations. What did you do with the drained coolant?

____

____

____

____

____

9. Use a knife to cut off the old hose and loosen or cut the old clamp. Task Completed ☐

10. Slide the old hose off the fitting. If the hose is stuck, do not pry it off; instead, cut it off. Task Completed ☐

11. Clean off any remaining hose particles with a wire brush or emery cloth. What can happen if dirt or metal burrs are on the fitting when a new hose is attached?

____

____

12. Coat the surface with a sealing compound. What compound did you use?

____

13. Place the new clamps on each end of the hose before positioning the hose. Task Completed ☐

14. Slide the clamps to about 1/4 inch from the end of the hose after it is properly positioned on the fitting. Task Completed ☐

15. Tighten the clamp securely. Do not overtighten. Task Completed ☐

16. Refill the cooling system with coolant. Task Completed ☐

17. Run the engine until it is warm. Task Completed ☐

18. Bleed the system according to the manufacturer's recommendations. What did you do to bleed the system? Task Completed ☐

____

____

____

____

19. Turn off the engine and recheck the coolant level. Task Completed ☐

20. Retighten the heater and cooling system hose clamps. Task Completed ☐

Problems Encountered

Instructor's Comments

HEATING AND AIR CONDITIONING JOB SHEET 23

Servicing a Thermostat

Name ______________________ Station ______________ Date ______________

NATEF Correlation

This Job Sheet addresses the following NATEF task:

C.4. Inspect, test, and replace thermostat and gasket.

Objective

Upon completion of this job sheet, you will be able to inspect, test, and replace a thermostat and gasket.

Tools and Materials

Thermometer

Container for water

Heat source for heating the water

Basic hand tools

Protective Clothing

Goggles or safety glasses with side shields

Describe the vehicle being worked on:

Year ________ Make ____________________ Model ____________________

VIN ____________________ Engine type and size ____________________

PROCEDURE

1. Thoroughly inspect the area around the thermostat and its housing. Describe the result of that inspection.

   ____

   ____

   ____

2. There are several ways to test the opening temperature of a thermostat. The first method does not require that the thermostat be removed from the engine. Begin by removing the radiator pressure cap from a cool radiator and inserting a thermometer into the coolant. Task Completed ☐

3. Start the engine and let it warm up. Watch the thermometer and the surface of the coolant. When the coolant begins to flow or move in the radiator, what is indicated?

   ____

   ____

4. When the fluid begins to flow, record the reading on the thermometer. If the engine is cold and coolant circulates, this means that the thermostat is stuck open and must be replaced. The measured opening temperature of the thermostat was ________________.

5. Compare the measured opening temperature with the specifications. The specifications call for the thermostat to open at what temperature? __________ What do you recommend based on the results of this test?

   ____

   ____

6. The other way to test a thermostat is to remove it. If the cap is still on, begin removal by opening the radiator pressure cap to relieve pressure. Task Completed ☐

7. Drain coolant from the radiator until the level of coolant is below the thermostat housing. Recycle the coolant according to local regulations. What did you do with the drained coolant?

   ____

   ____

   ____

8. Disconnect the upper radiator hose from the thermostat housing. Task Completed ☐

9. Unbolt and remove the housing from the engine. The thermostat may come off with the housing. Task Completed ☐

10. Thoroughly clean the gasket surfaces for the housing. Make sure the surfaces are not damaged while doing so and that gasket pieces do not fall into the engine at the thermostat bore. Task Completed ☐

11. Carefully inspect the thermostat housing. Describe your findings.

    ____

    ____

12. Suspend the thermostat while it is completely submerged in a small container of water. Make sure it does not touch the bottom of the container. What did you use to suspend it?

    ____

    ____

    ____

13. Place a thermometer in the water so that it does not touch the container and only measures water temperature. Task Completed ☐

14. Heat the water. When the thermostat valve barely begins to open, read the thermometer. What was the measured opening temperature of the thermostat? Compare that with the specifications.

    ____

    ____

15. Remove the thermostat from the water and observe the valve. Record what happened, as well as your conclusions about the thermostat.

____

____

____

16. Carefully look over the new thermostat (or the old one if it is still usable). Identify which end of the thermostat should face the radiator. How did you determine the proper direction for installation?

____

____

17. Fit the thermostat in the recessed area in the engine or thermostat housing. Where was the recess?

____

18. Refer to the installation instructions on the gasket's container. Should an adhesive or sealant, or both, be used with the gasket? If so, which should be used?

____

____

19. Install the gasket according to the instructions. Task Completed ☐

20. Install the thermostat housing. Before tightening this into place, make sure it is fully seated and flush onto the engine. Failure to do this will result in a broken housing. Task Completed ☐

21. Tighten the bolts evenly and carefully and to the correct specifications. What are the specifications?

____

22. Connect the upper radiator hose to the housing with a new clamp. Why should you use a new clamp?

____

____

____

23. Pressurize the system and check for leaks. Why should you do this now?

____

____

24. Replenish the coolant and bring it to its proper level. Install the radiator cap. Task Completed ☐

25. Run the engine until it is at normal operating temperature. Did the cooling system warm up properly and does it seem the thermostat is working properly?

____

____

26. Recheck the coolant level. Task Completed ☐

Problems Encountered

____

____

____

Instructor's Comments

____

____

____

HEATING AND AIR CONDITIONING JOB SHEET 24

Checking a Mechanical Cooling Fan

Name ______________________ Station ______________ Date ______________

NATEF Correlation

This Job Sheet addresses the following NATEF task:

C.7. Inspect and test cooling fan, fan clutch fan shroud, and air dams; perform necessary action.

Objective

Upon completion of this job sheet, you will be able to inspect and test a cooling fan, fan clutch, fan shroud, and air dams.

Tools and Materials

Vehicle with a mechanical cooling fan

Thermometer

Ignition timing light

Tachometer

Protective Clothing

Goggles or safety glasses with side shields

Describe the vehicle being worked on:

Year ________ Make ______________________ Model ______________________

VIN ______________________ Engine type and size ______________________

PROCEDURE

1. Inspect the fan shroud and air dams. Make sure they are firmly mounted to the radiator and/or frame of the vehicle. Summarize your findings.

   ____

   ____

2. Check the straightness of the fan blades. Summarize your findings.

   ____

   ____

3. Check for missing or damaged attaching rivets. Check the attaching point for each fan blade. Summarize your findings.

____

____

4. Check the condition of the drive pulley for the fan. Summarize your findings.

____

____

5. If the fan is equipped with a clutch, continue this job sheet. Task Completed ☐

6. Use your hands to spin the fan. If the fan rotates more than twice, the clutch is bad. Summarize your findings.

____

____

7. Check for oil around the clutch. Oil indicates the viscous clutch is leaking and should be replaced. Summarize your findings.

____

____

8. Attach a thermometer to the engine side of the radiator. Task Completed ☐

9. Connect a timing light and tachometer to the engine. Task Completed ☐

10. Start the engine, then observe and record the reading on the thermometer when the engine is cold. Task Completed ☐

11. Aim the timing light at the fan blades; they should appear to being moving very slowly. Summarize your findings.

____

____

12. Put a large piece of cardboard across the radiator to block air flow. Watch the thermometer and the timing light flashes as the engine warms up. Summarize your findings.

____

____

13. Refer to the service manual to find the specified fan engagement temperature. What is it?

____

____

14. Once the engine reaches that temperature, remove the cardboard. The fan clutch should now cause the fan speed to increase. Summarize your findings.

____

____

Problems Encountered

____

____

____

Instructor's Comments

____

____

____

HEATING AND AIR CONDITIONING JOB SHEET 25

Clean, Inspect, Test, and Replace Electric Cooling Fans and Cooling System-Related Temperature Sensors

Name ______________________ Station ______________ Date ______________

NATEF Correlation

This Job Sheet addresses the following NATEF task:

C.8. Inspect and test electrical cooling fan, fan control system, and circuits; determine necessary action.

Objective

Upon completion of this job sheet, you will be able to demonstrate the ability to inspect, test, clean, and replace electric cooling fans and cooling system-related temperature sensors.

Tools and Materials

Hand tools

Test light (circuit tester)

Digital multimeter (DMM)

Jumper wire

Service manual

Protective Clothing

Goggles or safety glasses with side shields

Describe the vehicle being worked on:

Year __________ Make ____________________ Model ____________________

VIN ____________________ Engine type and size ____________________

Describe general condition:

____

PROCEDURE

1. Visually inspect the electric cooling fan(s) and related wiring to determine if they are clean and properly connected. Task Completed ☐
2. Run the engine until it reaches normal operating temperature to determine if the electric cooling fan is working. Task Completed ☐
3. If the fan is operating properly, blow the fan and surrounding area clean with air. Task Completed ☐
4. If the fan is not working, check the power wire to the fan to determine if there is power. Task Completed ☐

5. Check the ground circuit to determine if the system has a good ground. Task Completed ☐
6. If power and a good ground are present, remove the electric fan assembly and replace it. Task Completed ☐
7. If there is no power, check the fuse and fan relay. The location of the fuse and relay can be found in the service manual. Task Completed ☐
8. Replace bad components, then retest fan operation. Task Completed ☐
9. If there is not a good ground, check the temperature sensor that controls the electric fan. The location of the sensor and the proper test procedure can be found in the service manual. Task Completed ☐
10. If the sensor fails the test, replace it. Check the operation of the electric fan. Task Completed ☐

WARNING: *The electric cooling fans can operate with the engine and the key both turned off. The fan can come on at any time if the engine is hot or there is a problem with a sensor. Keep your hands away from the fan at all times.*

Problems Encountered

____

____

____

Instructor's Comments

____

____

____

HEATING AND AIR CONDITIONING JOB SHEET 26

Testing Heater Control Valves

Name ______________________ Station ______________ Date ______________

NATEF Correlation

This Job Sheet addresses the following NATEF task:

C.9. Inspect and test heater control valve(s); perform necessary action.

Objective

Upon completion of this job sheet, you will be able to inspect and test heater control valves properly.

Tools and Materials

Hand-operated vacuum pump

Protective Clothing

Goggles or safety glasses with side shields

Describe the vehicle being worked on:

Year ________ Make ____________________ Model ____________________

VIN ____________________ Engine type and size ____________________

PROCEDURE

1. Locate the heater control valve and inspect the hoses' connections and their clamps. Summarize your findings.

   ____

   ____

2. If the valve is mechanically controlled, check the condition of the linkage and the cable. Summarize your findings.

   ____

   ____

3. If the valve is vacuum controlled, check the condition of the vacuum hose and connection at the valve. Summarize your findings.

   ____

   ____

4. Disconnect the vacuum hose and connect a hand-operated vacuum pump to the valve. Task Completed ☐

5. Pump a vacuum at the valve and watch its operation; it should open smoothly. Summarize your findings.

____

____

6. Refer to the service manual to determine if the valve is normally open or closed. Turn the heater control on the instrument panel off. Check the valve to see if it is where it should be. Summarize your findings.

____

____

7. Move the heater control to full heat and check the position of the valve. Summarize your findings.

____

____

Problems Encountered

____

____

____

Instructor's Comments

____

____

____

HEATING AND AIR CONDITIONING JOB SHEET 27

Removing and Installing a Heater Core

Name ______________________ Station ______________ Date ______________

NATEF Correlation

This Job Sheet addresses the following NATEF task:

C.10. Remove, inspect, and reinstall heater core.

Objective

Upon completion of this job sheet, you will be able to correctly remove and reinstall a heater core.

Tools and Materials

Catch basin   Coolant recycling machine

Hand tools   Service manual

Protective Clothing

Goggles or safety glasses with side shields

Describe the vehicle being worked on:

Year ________ Make __________________ Model ________________

VIN __________________ Engine type and size ____________________

PROCEDURE

1. Drain and recycle the coolant from the engine with the approved recycling equipment. Task Completed ☐
2. Disconnect the negative battery cable. Task Completed ☐
3. Refer to the service manual for the recommended procedure for removing the heater core. Summarize the steps here:

   ____

   ____

   ____

   ____

4. If the evaporator and heater core are a combined unit, recover and recycle the refrigerant in the A/C system. Task Completed ☐
5. Disconnect and label all electrical connectors that are connected to the heater core and all those that may be in the way. What did you need to disconnect?

   ____

   ____

6. Loosen the hose clamps on the hoses connected to the heater core. Task Completed ☐

7. Remove the coolant hoses at the heater core and plug or cap the hose ends to prevent more lose of coolant. Task Completed ☐

8. Inspect the heater hoses and record your findings here:

____

____

9. Replace the heater hoses if necessary or desired. Task Completed ☐

10. Unbolt and remove the heater core. Task Completed ☐

11. Loosely place the new heater core into position. Task Completed ☐

12. Install new hose clamps over the hose ends. Task Completed ☐

13. Slip the hose ends into position over the inlet and outlet nipples on the heater core. Task Completed ☐

14. Tighten the hose clamps. Task Completed ☐

15. Tighten the mounting bolts for the heater core. Task Completed ☐

16. Reinstall and reconnect all parts, wires, cables, and vacuum hoses that were disconnected or removed during removal. Task Completed ☐

17. Fill the cooling system with new clean coolant and bleed the system. Task Completed ☐

18. Connect the negative battery cable. Task Completed ☐

19. Perform a leak test and correct any problems. Were there any problems?

____

____

20. Drive the vehicle and check the heating system. Does the system work as it should? Explain.

____

____

____

____

Problems Encountered

____

____

____

Instructor's Comments

____

____

____

HEATING AND AIR CONDITIONING JOB SHEET 28

Testing HVAC Controls

Name ______________________________ Station __________________ Date __________________

NATEF Correlation

This Job Sheet addresses the following NATEF tasks:

D.1. Diagnose malfunctions in the electrical controls of heating, ventilation, and A/C (HVAC) systems; determine necessary action.

D.4. Diagnose malfunctions in the vacuum, mechanical, and electrical components and controls of the heating, ventilation, and A/C (HVAC) system; determine necessary action.

D.6. Inspect and test A/C-heater control cables, motors, and linkages; perform necessary action.

Objective

Upon completion of this job sheet, you will be able to diagnose failures in the electrical, vacuum, and mechanical controls of heating, ventilation, and air conditioning systems.

Tools and Materials

Service manual

Vacuum gauge

Hand-operated vacuum pump

Protective Clothing

Goggles or safety glasses with side shields

Describe the vehicle being worked on:

Year ___________ Make __________________________ Model __________________________

VIN __________________________ Engine type and size __________________________

PROCEDURE

1. Check the fuses for the compressor clutch and the blower motor. Summarize your findings.

____

____

2. Check the operation of the blower switch in all switch positions. Summarize your findings.

____

____

3. If there is a problem with the blower motor, check the circuit to determine the exact fault. Then, summarize your findings.

____

____

4. With the engine running, turn the A/C on and off and listen for the engagement of the compressor clutch. If the clutch makes a chattering noise, it may need adjustment or replacement. Summarize your findings.

____

____

5. If the clutch did not cycle with the switch, check to see if the clutch is constantly engaged or disengaged. If always engaged, the clutch is bad or there is a wire-to-wire short providing voltage to the clutch at all times. If the clutch is always disengaged, there is probably a problem with the clutch control circuit. Summarize your findings.

____

____

6. Check the electrical connections at the clutch. Summarize your findings.

____

____

7. Connect a jumper wire from the battery to the clutch. Check its operation now. If it operates quietly and smoothly, the problem is in the clutch control circuit. If it didn't operate normally when jumped, the clutch is bad. Summarize your findings.

____

____

8. Check the resistance of the clutch coil with an ohmmeter. Compare your measurement to specifications. Summarize your findings.

____

____

9. Using the service manual, identify any sensors or switches that control or limit the operation of the clutch. List all of them and summarize their condition.

____

____

____

____

10. Identify and inspect the hoses and components for vacuum motors and doors. Look for disconnected or broken hoses, broken connectors, misrouted vacuum lines, and loose or disconnected electrical connectors at vacuum switches and controls. Summarize your findings.

11. Check the vacuum at the engine's intake manifold. Do this while the engine is idling. Summarize your findings.

12. Check the individual vacuum components. Use a hand-operated vacuum pump to check the component's activity and its ability to hold a vacuum. Summarize your findings.

13. Inspect all mechanical or cable linkages and controls. Make sure the cables are properly attached to the levers of the control switch. Summarize your findings.

14. If the cable(s) is equipped with an automatic adjuster, make sure the adjusting mechanism operates freely and is not damaged. Summarize your findings.

Problems Encountered

Instructor's Comments

HEATING AND AIR CONDITIONING JOB SHEET 29

Diagnosing Motor-Driven Accessories

Name ______________________ Station ______________ Date ______________

NATEF Correlation

This Job Sheet addresses the following NATEF task:

D.2. Inspect and test A/C-heater blower, motors, resistors, switches, relays, wiring, and protection devices; perform necessary action.

Objective

Upon completion of this job sheet, you will be able to inspect and test A/C-heater blower motor circuits and components.

Tools and Materials

Wiring diagram for the vehicle

DMM

Protective Clothing

Goggles or safety glasses with side shields

Describe the vehicle being worked on:

Year ________ Make ____________________ Model ____________________

VIN ____________________ Engine type and size ____________________

PROCEDURE

NOTE: *Although this job sheet is focused on the blower motor, the procedures can easily be applied to all motor-driven accessories. Refer to the appropriate service manual and wiring diagram to identify the components of the circuit you wish to test. Then apply the same sequence and logic given in this job sheet.*

1. Refer to the service or owner's manual to identify the number of speeds at which the blower motor should operate. Also, identify whether the motor should run when the switch is in the off position or its lowest position. Describe how the blower motor should operate.

   ____

   ____

2. Check the wiring diagram and identify whether the blower circuit is controlled by a ground side switch or an insulated (power side) switch. State which one.

   ____

   ____

3. Start the engine and turn on the blower. Move the blower control to all available positions and summarize what happened in each.

____

____

4. Match the type of circuit control and the problem with one of the following. Task Completed ☐

The blower works at some speeds but not all.

a. Check the voltage to the blower motor at the various switch positions. Task Completed ☐

b. If the voltage doesn't change when a new position is selected, check the circuit from the switch to the resistor block. Task Completed ☐

c. If there was zero voltage in a switch position, check for an open in the resistor block. Task Completed ☐

d. Give a summary of the test results.

____

____

The blower doesn't work and is controlled by an insulated switch.

a. Check the fuse or circuit breaker. If it is open, check the circuit for a short. Task Completed ☐

b. Connect a jumper wire from a power source to the motor. If the motor operates, there is an open in the circuit between the fuse and the motor. Task Completed ☐

c. To identify the location of the open, connect a jumper wire across the switch. If the motor now operates, the switch is bad. Task Completed ☐

d. Connect a jumper wire across the resistor block. If the motor now operates, the resistor block is bad. Task Completed ☐

e. If the motor did not operate when the jumper wire connected it to a power source, connect the jumper wire from the motor to a known good ground. If the motor operates, the fault is in the ground circuit. If the motor doesn't operate, the motor is bad. Task Completed ☐

f. Give a summary of the test results.

____

____

The blower doesn't work and is controlled by a ground side switch.

a. Check the fuse or circuit breaker. If it is open, check the circuit for a short. Task Completed ☐

b. Connect a jumper wire from the motor's ground terminal to a known good ground. If the motor runs, the problem is in that circuit. Task Completed ☐

c If the motor did not operate, check for voltage at the battery terminal of the motor. If no voltage is found, there is an open in the power feed circuit for the motor. If voltage was present, the motor is bad. Task Completed ☐

d. If the motor operated when the jumper wire was connected in step b, use an ohmmeter to check the ground connection of the switch. Task Completed ☐

e. If the ground is good, use a voltmeter to probe for voltage at any of the circuits from the resistor block to the switch. Replace the switch if there is power at that point. Task Completed ☐

f. If no voltage was available at the switch, check the circuit for an open. If there is not an open, suspect a faulty switch. Task Completed ☐

g. Give a summary of the test results.

___

___

The blower fan runs constantly.

a. If the motor runs when it should not, there is probably a short in the circuit. If a ground side switch controls the circuit, check for a short to ground in the control circuit. The exact problem can be isolated by disconnecting portions of the circuit until the motor stops. The short is in the part of the circuit that disconnected last. Task Completed ☐

b. If the circuit is controlled by an insulated switch, check for a wire-to-wire short. Check other circuits of the vehicle to identify what circuit is involved in this problem. That circuit will also experience a lack of control, or when that circuit is turned off, the blower motor will turn off. The exact problem can be isolated by disconnecting portions of the circuit until the motor stops. The short is in the part of the circuit that disconnected last. Task Completed ☐

c. Give a summary of the test results.

___

___

Problems Encountered

___

___

___

Instructor's Comments

___

___

___

HEATING AND AIR CONDITIONING JOB SHEET 30

Servicing Compressor Load Sensors

Name ______________________ Station ______________ Date ______________

NATEF Correlation

This Job Sheet addresses the following NATEF task:

D.3. Test and diagnose A/C compressor clutch control systems; determine necessary action.

Objective

Upon completion of this job sheet, you will be able to demonstrate the ability to accurately test A/C compressor load cutoff systems.

Tools and Materials

Hand tools

DMM

Service manual

Protective Clothing

Goggles or safety glasses with side shields

Describe the vehicle being worked on:

Year ________ Make ____________________ Model ____________________

VIN ____________________ Engine type and size ____________________

PROCEDURE

1. Using the service manual, locate the various pressure and temperature sensors in the system that control the operation of the compressor. List them and describe their location.

   ____

   ____

   ____

   ____

2. Check the manual to determine if each of these is normally closed or normally open, and summarize your findings.

   ____

   ____

   ____

3. Checking one sensor at a time, complete the following steps.
   a. Disconnect the wires at the sensor or switch.
   b. Connect the DMM across the switch terminals.
   c. Set the meter for resistance or continuity checks.
   d. If the switch is normally closed, the reading should be zero ohms.
   e. If the switch is normally open, you should get an infinite reading.
   f. Any reading other than these is an indication of a faulty switch.
   g. Summarize your results from each switch.

____

____

____

____

____

4. If a switch needs to be replaced, the system may need to be discharged. Check the service manual before proceeding. Which switches can be replaced without discharging the refrigerant in the system?

____

____

____

____

5. To remove the defective switch, loosen it and remove it. Task Completed ☐
6. To install the new switch, carefully thread it into place and tighten it. Task Completed ☐
7. Reconnect the wires leading to the switches. Task Completed ☐

Problems Encountered

____

____

____

Instructor's Comments

____

____

____

HEATING AND AIR CONDITIONING JOB SHEET 31

Checking an A/C Master Control Unit

Name ______________________ Station ______________ Date ______________

NATEF Correlation

This Job Sheet addresses the following NATEF task:

D.5. Inspect and test A/C-heater control panel assembly; determine necessary action.

Objective

Upon completion of this job sheet, you will be able to inspect and test an A/C-heater control panel assembly.

Tools and Materials

DMM

Service manual

Hand-operated vacuum pump

Protective Clothing

Goggles or safety glasses with side shields

Describe the vehicle being worked on:

Year ________ Make ____________________ Model ____________________

VIN ____________________ Engine type and size ____________________

PROCEDURE

1. Remove the A/C-heater master control from the instrument panel. Task Completed ☐
2. Inspect the connectors that go to the control unit and record your findings. Look for damage and corrosion.

   ____

   ____

3. Thoroughly inspect the connectors at the control unit. Look for terminal distortion (bends), corrosion, and other damage. Record your findings.

   ____

   ____

4. With an ohmmeter, check the continuity through the blower motor switch in all its positions. Summarize your findings.

   ____

   ____

5. Compare your results with the wiring diagram for the switch. Do you have good continuity at the correct terminals when the switch is in its various positions? _____________

6. With the ohmmeter, check the resistance across the temperature control switch. Watch the meter as you sweep through the full range of the switch, and record your findings.

____

____

____

7. If the mode selector switch is electrical, check for continuity across the terminals as you move the switch through the various mode selections. Record your findings.

____

____

____

8. If the mode selector switch is a vacuum switch, apply a vacuum to the inlet of the switch and feel for a vacuum at the various tube connectors on the switch while you move the selector through its various positions. Record your findings.

____

____

____

9. Compare your results to the information in the service manual, and summarize your recommendations for service to the master control unit.

____

____

____

____

10. Reinstall the old switch or install a new one. Task Completed ☐

Problems Encountered

____

____

____

Instructor's Comments

____

____

____

HEATING AND AIR CONDITIONING JOB SHEET 32

Inspecting the Air Duct System

Name ______________________ Station ______________ Date ______________

NATEF Correlation

This Job Sheet addresses the following NATEF task:

D.7. Inspect and test A/C-heater ducts, doors, hoses, cabin filters, and outlets; perform necessary action.

Objective

Upon completion of this job sheet, you will be able to inspect and test the A/C-heater ducts, doors, hoses, and outlets.

Tools and Materials

Service manual

Vacuum pump

Vacuum gauge

Hand tools

Protective Clothing

Goggles or safety glasses with side shields

Describe the vehicle being worked on:

Year ________ Make ____________________ Model ____________________

VIN ____________________ Engine type and size ____________________

PROCEDURE

1. Identify the vacuum source for the various doors and switches of the duct system and describe its location.

   ____

   ____

2. Disconnect that vacuum source. Task Completed ☐

3. Connect the vacuum pump to the vacuum reserve tank.

4. Check for available vacuum at each of the vacuum-operated doors, motors, and switches. Summarize your findings.

   ____

   ____

5. Connect the vacuum gauge to the inlet for the defroster door motor(s) and record the vacuum available there when the master control switch is in the following positions (there are many variations in the names used for these operating modes; use the one that best describes the position you are testing).

   a. MAX ________ in. Hg
   b. NORM ________ in. Hg
   c. Bilevel ________ in. Hg
   d. VENT ________ in. Hg
   e. HEAT ________ in. Hg
   f. BLEND ________ in. Hg
   g. DEFROST ________ in. Hg
   h. OFF ________ in. Hg

6. Connect the vacuum gauge to the inlet for the A/C door motor(s) and record the vacuum available there when the master control switch is in the following positions.

   a. MAX ________ in. Hg
   b. NORM ________ in. Hg
   c. Bilevel ________ in. Hg
   d. VENT ________ in. Hg
   e. HEAT ________ in. Hg
   f. BLEND ________ in. Hg
   g. DEFROST ________ in. Hg
   h. OFF ________ in. Hg

7. Connect the vacuum gauge to the inlet for the outside (vent) door motor(s), and record the vacuum available there when the master control switch is in the following positions.

   a. MAX ________ in. Hg
   b. NORM ________ in. Hg
   c. Bilevel ________ in. Hg
   d. VENT ________ in. Hg
   e. HEAT ________ in. Hg
   f. BLEND ________ in. Hg
   g. DEFROST ________ in. Hg
   h. OFF ________ in. Hg

8. Connect the vacuum gauge to the inlet for the bilevel door motor(s) and record the vacuum available there when the master control switch is in the following positions.

   a. MAX ________ in. Hg
   b. NORM ________ in. Hg
   c. Bilevel ________ in. Hg
   d. VENT ________ in. Hg
   e. HEAT ________ in. Hg
   f. BLEND ________ in. Hg
   g. DEFROST ________ in. Hg
   h. OFF ________ in. Hg

9. Compare your findings to the information given in the service manual, and state your conclusions and recommendations for system service.

____

____

____

____

10. The in-cabin dust and pollen filter should be replaced at a fixed interval, normally every 30,000 miles (48,000 km) or 24 months. When is it recommended that the filter be changed on this vehicle?

____

11. With the blower on high speed, feel the airflow from the ducts and vents. If the flow seems to be less than it should, replace the filter. What did you find?

____

____

12. Normally to replace the filter, the glove box needs to be removed to access the filter's holder. Does the glove box need to be removed on this vehicle?

____

Problems Encountered

____

____

____

Instructor's Comments

____

____

____

HEATING AND AIR CONDITIONING JOB SHEET 33

Checking Automatic Climate Control Systems

Name ______________________ Station ______________ Date ______________

NATEF Correlation

This Job Sheet addresses the following NATEF task:

D.8. Check operation of automatic and semiautomatic HVAC control systems; determine necessary action.

Objective

Upon completion of this job sheet, you will be able to check the operation of automatic and semiautomatic HVAC control systems.

Tools and Materials

Scan tool

DMM

Service manual

Protective Clothing

Goggles or safety glasses with side shields

Describe the vehicle being worked on:

Year ________ Make ____________________ Model ____________________

VIN ____________________ Engine type and size ____________________

PROCEDURE

1. Refer to the service manual and determine if the automatic climate control system relies on the PCM, body control module (BCM), or a separate computer for control. What did you find?

   ____

   ____

2. If the unit is part of the PCM or BCM system, connect the scan tool and retrieve any codes that may be present. What did you find?

   ____

   ____

3. If the unit is controlled by its own computer system, use your service manual and record the procedure for retrieving trouble codes.

   ____

   ____

   ____

4. If there were trouble codes, what problem areas were identified?

____

____

5. Test the components or subsystems identified by the trouble codes, and summarize the results of those tests.

____

____

6. If the system is a semiautomatic system, check the service manual for the proper diagnostic procedures for checking evaporator and heating controls. Record these procedures here.

____

____

____

7. Follow the appropriate test procedures and summarize your findings.

____

____

____

____

____

Problems Encountered

____

____

____

Instructor's Comments

____

____

____

HEATING AND AIR CONDITIONING JOB SHEET 34

Checking the Operation of Refrigerant Handling Equipment

Name ______________________ Station ______________ Date ______________

NATEF Correlation

This Job Sheet addresses the following NATEF task:

E.1. Perform correct use and maintenance of refrigerant handling equipment.

Objective

Upon completion of this job sheet, you will be able to verify correct operation and maintenance of refrigerant handling equipment.

Tools and Materials

Manifold gauge sets

Refrigerant storage containers

Refrigerant recovery machines

Protective Clothing

Goggles or safety glasses with side shields

PROCEDURE

1. Check the manifold gauge set to make sure it has a cut off valve at the end of each hose so that the fitting not in use is automatically shut. Summarize your findings.

   ____

2. Check to make sure the manifold gauge set is clearly labeled as to what refrigerant it should be used for. Manifold gauge sets for R-134a can be identified by one or all of the following: Labeled FOR USE WITH R-134a, Labeled HFC-134 or R-134a, and/or have a light blue color on the face of the gauges. Summarize your findings.

   ____

3. Verify that the R-134a service hoses have a black stripe along their length and are clearly labeled SAE J2196/R-134a. The low pressure hose is blue with a black stripe, the high pressure hose is red with a black stripe, and the center service hose is yellow with a black stripe. Summarize your findings.

   ____

   ____

4. Check the fittings at the end of the service hoses. By law, the service hoses for one type of refrigerant should not easily connect into the wrong system, as the fittings for an R-134a system are different from those used in an R-12 system. Summarize your findings.

____

____

5. Make sure all A/C equipment is designated as to what refrigerant it is designed for. Because R-134a is not interchangeable with R-12, separate sets of hoses, gauges, and other equipment are required to service vehicles. All equipment used to service R-134a and R-12 systems must meet SAE standard J1991. Summarize your findings.

____

6. Make sure the containers of refrigerant are clearly labeled. When you are not sure of the refrigerant stored in a container or if you suspect a mixing of refrigerants has occurred, you should run a purity test and/or use a refrigerant identifier. Summarize your findings.

____

____

7. Check the refrigerant storage containers. All recycled refrigerant must be safely stored in DOT CFR Title 49 or UL-approved containers. Containers specifically made for R-134a should be so marked. Before any container of recycled refrigerant can be used, it must be checked for noncondensable gases. Summarize your findings.

____

____

8. Locate the UL approved label on the refrigerant recovery machines. Summarize your findings.

____

____

Problems Encountered

____

____

____

Instructor's Comments

____

____

____

HEATING AND AIR CONDITIONING JOB SHEET 35

Identify and Recover Refrigerant

Name ______________________ Station ______________ Date ______________

NATEF Correlation

This Job Sheet addresses the following NATEF task:

E.2. Identify (by label application or use of a refrigerant identifier) and recover A/C system refrigerant.

Objective

Upon completion of this job sheet, you will be able to identify and recover refrigerant.

Tools and Materials

Fender covers

Manifold gauge set

Service manual

Protective clothing

Goggles or safety glasses with side shields

Describe the vehicle being worked on:

Year ________ Make ____________________ Model ____________________

VIN ____________________ Engine type and size ____________________

Describe the type of air conditioning found on the vehicle:

____

PROCEDURE

1. Place fender covers over the front fenders for protection. Task Completed ☐
2. Check the decal under the hood on the air-conditioning unit for the type of freon used in your specific system. Task Completed ☐
3. If this decal is not present, refer to the service manual for your specific vehicle. Find the type of freon used in your system. Task Completed ☐
4. One other way to determine if the system uses R-12 or R-134a is to look at the size of the Schrader valves. If the valves are smaller, about the size of a tire Schrader valve, the system uses R-12. If the valves are considerably larger than a tire Schrader valve, the system would most likely use R-134a. Task Completed ☐

   NOTE: *It is very important to determine the type of refrigerant in your system before servicing. These two types of refrigerant cannot be mixed.*

5. Connect the manifold gauge set to the vehicle. Task Completed ☐

6. Connect the center hose of the manifold set to the recovery unit. Task Completed ☐

7. Begin the recovery procedure following the instructions for your specific vehicle. Task Completed ☐

8. When the recovery is complete, close the valves on the manifold gauge set and turn off the recovery unit. Task Completed ☐

9. You are now prepared to do necessary repairs on the air-conditioning system. Task Completed ☐

Problems Encountered

___

___

___

Instructor's Comments

___

___

___

HEATING AND AIR CONDITIONING JOB SHEET 36

Recycle, Label, and Store Refrigerant

Name ______________________ Station ______________ Date ______________

NATEF Correlation

This Job Sheet addresses the following NATEF tasks:

E.3. Recycle refrigerant.

E.4. Label and store refrigerant.

Objective

Upon completion of this job sheet, you will be able to recycle, label, and store refrigerant safely and properly.

Tools and Materials

Recycling machine, Service manual

Protective Clothing

Goggles or safety glasses with side shields

Describe the vehicle being worked on:

Year ________ Make ____________________ Model ____________________

VIN ____________________ Engine type and size ____________________

Describe the type of air conditioning found on the vehicle:

____

PROCEDURE

1. Refer to the manual on your specific machine to determine the procedure to recycle refrigerant. Task Completed ☐
2. Isolate recycling machine from the vehicle system. Task Completed ☐
3. Recycle the refrigerant. Task Completed ☐
4. Label the container as "empty" or "empty, evacuated and ready for disposal" or "freon" or "contains recycled freon." Task Completed ☐
5. After the freon container has been identified, it can be stored or disposed of accordingly. Task Completed ☐

Problems Encountered

___

___

___

Instructor's Comments

___

___

___

HEATING AND AIR CONDITIONING JOB SHEET 37

Testing for Noncondensable Gases

Name ________________________ Station ________________ Date ________________

NATEF Correlation

This Job Sheet addresses the following NATEF task:

E.5. Test recycled refrigerant for noncondensable gases.

Objective

Upon completion of this job sheet, you will be able to test recycled refrigerant for noncondensable gases according to the guidelines set by the SAE.

Tools and Materials

Pressure gauge set

Thermometer

SAE pressure/temperature chart

Protective Clothing

Goggles or safety glasses with side shields

Describe the vehicle being worked on:

Year __________ Make ______________________ Model ______________________

VIN ______________________ Engine type and size ______________________

PROCEDURE

Task Completed ☐

1. To determine if the recycled refrigerant container has excess noncondensable gases, the container must be stored at a temperature of 65°F (18.3°C) or above for 12 hours and protected from the sun.
2. Install a calibrated pressure gauge onto the container and measure the pressure in the container. What pressure was read?

   ____

3. Measure the temperature of the air 4 inches away from the container's surface. What was your measurement?

   ____

4. Compare the measured pressure and temperature to the pressure/temperature chart. Determine if the recycled refrigerant has excessive noncondensable gases. Explain your findings.

   ____

   ____

   ____

5. If the refrigerant has excessive noncondensable gases, it should be recycled again and then retested before it is used. Task Completed ☐

6. Summarize the results of this test.

____

____

____

Problems Encountered

____

____

____

Instructor's Comments

____

____

____

HEATING AND AIR CONDITIONING JOB SHEET 38

Evacuating and Charging an A/C System

Name ______________________ Station ______________ Date ______________

NATEF Correlation

This Job Sheet addresses the following NATEF task:

E.6. Evacuate and charge an A/C system.

Objective

Upon completion of this job sheet, you will be able to evacuate and recharge an A/C system with a manifold gauge set and a recycling and charging station.

Tools and Materials

Manifold gauge set

Refrigerant recycling/reclaiming machine

Vacuum pump

Proper adapters for the fittings

Protective Clothing

Goggles or safety glasses with side shields

Describe the vehicle being worked on:

Year ________ Make ___________________ Model ___________________

VIN ___________________ Engine type and size ___________________

Type of refrigerant used in the system: __________

PROCEDURE

Using a Manifold Gauge Set

1. Make sure the refrigerant has been removed from the system with a refrigerant reclaiming machine. Do not vent the refrigerant into the atmosphere. Task Completed ☐
2. Close the hand valves on the gauge set. Task Completed ☐
3. Connect the hoses for manifold gauge set to the high-side and low-side service ports. Task Completed ☐
4. If the system is equipped with shutoff type service valves, open them slightly. Task Completed ☐
5. Connect the center manifold service hose to the intake of the vacuum pump. Task Completed ☐
6. Open the shutoff valves at each of the three service hoses. Task Completed ☐

7. Turn on the vacuum pump. Task Completed ☐
8. Open the low-side manifold hand valve and watch the low-side gauge. Did it drop to indicate a slight vacuum?

____

____

9. Once the gauge reads about 25 in. Hg, close the low-side hand valve and watch the gauge reading. Is the system holding the vacuum? If not, what is indicated?

____

____

10. Reopen the low-side hand valve. Task Completed ☐
11. Open the high-side hand valve. Task Completed ☐
12. Allow the vacuum pump to run until close to 30 in. Hg is showing on the low-side gauge. About how long did this take?

____

____

13. Close both hand valves. Task Completed ☐
14. Refer to the service manual and identify the proper type and quantity of refrigerant to use in this system and record this here.

____

____

15. Will you be using pound cans or a larger can of refrigerant to charge the system?

____

16. Make sure the shutoff valves, manifold hand valves, and compressor service valves are closed. Task Completed ☐
17. Connect the center manifold service hose to the container of refrigerant. Task Completed ☐
18. Open the valve on the container. Task Completed ☐
19. Open the service hose, high-side, and low-side shutoff valves. Task Completed ☐
20. Start the engine and run it at a fast idle. Task Completed ☐
21. Turn on the A/C and set it and its blower to its maximum position. Task Completed ☐
22. Open the low-side manifold hand valve. Task Completed ☐
23. Allow only the recommended amount of refrigerant to enter the system. Task Completed ☐
24. Once the system is fully charged, close the low-side manifold hand valve. Task Completed ☐
25. Close the service hose shutoff valve. Task Completed ☐

26. Disconnect the service hose from the container. Task Completed ☐
27. Observe the pressures on the gauges. What do they indicate?

____

____

28. Return the engine to a normal idle speed and turn the A/C system off. Task Completed ☐
29. Turn off the engine. Task Completed ☐
30. Close all valves and use a recovery system to remove any refrigerant from the manifold and its hoses. Task Completed ☐
31. Remove the manifold gauge set. Task Completed ☐
32. Replace all protective caps and covers. Task Completed ☐

Using a Recycling and Charging Station

1. Physically locate the pressure fittings of the system you will be working with. Describe their location.

____

____

2. Make sure you have the proper adapters for the fittings prior to starting any service work on air conditioning systems. Task Completed ☐
3. Connect a gauge set to the system. Task Completed ☐
4. Connect the reclaiming unit to the gauge set. Task Completed ☐
5. Activate the reclaimer to discharge the system, and then begin evacuation. Allow enough time for the machine to draw a good vacuum (at least 29 in. Hg.) in the system. Observe the gauge set to determine when a vacuum is created. How long was it necessary to pump down the system?

____

6. Stop evacuation, and let the system stay with a vacuum for at least ten minutes. Check the vacuum at that time. Did the vacuum stay? What does this indicate?

____

____

____

7. Close the valves at the gauge set and disconnect the reclaiming machine. Task Completed ☐
8. Connect the charging station and open the gauge set valves. Task Completed ☐
9. Set the unit to deliver the required amount of refrigerant for that system. What is the proper charge?

____

10. After the system has been recharged, check the gauge readings while the system is in operation to make sure it is working properly. The gauge readings are:

____

____

____

____

Problems Encountered

____

____

____

Instructor's Comments

____

____

____